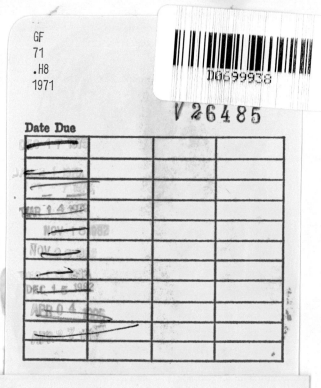

CIVILIZATION AND CLIMATE

CIVILIZATION

AND CLIMATE

BY

Ellsworth Huntington

Third Edition
Revised and Enlarged

ARCHON BOOKS

1971

Reprinted as an Archon Book 1971 with permission in
an unaltered and unabridged edition from the fifth
printing of the third edition of 1925

The Shoe String Press, Inc., Hamden, Connecticut 06514

ISBN 0-208-01159-5
Library of Congress catalog card number 78-143891

Printed in the United States of America

TABLE OF CONTENTS

LIST OF ILLUSTRATIONS

PREFACE TO FIRST EDITION

THIS volume is a product of the new science of geography. The old geography strove primarily to produce exact maps of the physical features of the earth's surface. The new goes farther. It adds to the physical maps an almost innumerable series showing the distribution of plants, animals, and man, and of every phase of the life of these organisms. It does this, not as an end in itself, but for the purpose of comparing the physical and organic maps and thus determining how far vital phenomena depend upon geographic environment. Among the things to be mapped, human character as expressed in civilization is one of the most interesting and one whose distribution most needs explanation. The only way to explain it is to ascertain the effect of each of many coöperating factors. Such matters as race, religion, institutions, and the influence of men of genius must be considered on the one hand, and geographical location, topography, soil, climate, and similar physical conditions on the other. This book sets aside the other factors, except incidentally, and confines itself to climate. In that lie both its strength and weakness. When the volume was first planned, I contemplated a discussion of all the factors and an attempt to assign to each its proper weight. The first friend whom I consulted advised a directly opposite course, whereby the emphasis should be centered upon the new climatic facts which seem to afford ground for a revision of some of our old estimates of the relation between man and his environment. In writing the book I have growingly felt the wisdom of that advice, and have been impressed with the importance of concentration upon a single point, even at the expense of seeming to take a one-sided view.

If the reader feels that due weight is not given to one factor or another, he must remember that many unmentioned phases of the subject have been deliberately omitted to permit fuller emphasis upon the apparent connection between a stimulating climate and high civilization.

The materials for this volume have been derived from a great variety of sources. Although personal observation and investigation are the basis of much that is here stated, still more has been derived from the world's general store of knowledge. Except in a few special cases I have not attempted to give references. To the general reader footnotes are not only useless, but often a distraction and nuisance. The careful student, on the other hand, cannot form a fair estimate of the hypothesis here presented without reading previous publications in which I have set forth the reasons for many conclusions which are not fully discussed here for lack of space. These publications contain numerous references. Accordingly, the needs of the student will be met by giving a brief list of books and articles which have served as preliminary steps to the present volume. These publications form a logical series with only such repetition as is necessary to make each a complete unit. It is scarcely necessary to add that the rapid growth of the subject during the past ten years has led to important modifications in some of the earlier conclusions.

The facts set forth in this volume have by no means been derived wholly from observation and reading. Not far from a hundred people have given direct personal assistance. They are so numerous that it is impossible to mention them all by name. Therefore it seems best not to single out any for special thanks. Many of my colleagues among the Yale faculty and among the geographers of America have gone out of their way to offer suggestions, or friendly criticisms, or to bring to my attention publications and facts that might have escaped my notice. Others connected with such organizations as the Carnegie Institution of Washington and the United States Weather Bureau have placed me under obligation by the kind way in which they have taken a personal rather than official interest in answering queries and providing data. Equally great courtesy has been shown by officers and other members of the teaching force at West Point and Annapolis, and by officials connected with various factories, including some whose figures it has not yet been possible to tabulate. Another large group comprises contributors to the map of civilization, many of whom devoted to this work time which they could ill afford. Lastly, I owe much to personal friends who fall in none of the groups already specified. I suppose that the total time given to this book by all these scores of people makes their contribution larger than mine. My chief hope is that they may feel that their kindness has not been wasted. To each and all I can only express my deep sense of gratitude, and most of all to those whose advice from the beginning has done more than anything else to keep this book true to a single aim.

E. H.

Yale University,
New Haven, Conn.,
July, 1915.

PREFACE TO THIRD EDITION

IN preparing the third edition of this book two circumstances have led to a radical revision. First, the World War has receded far enough into the background so that people's minds are once more able to concentrate upon the far-reaching problems of science, instead of the temporary details associated with war. During the past three years this has led to a marked increase in the number of publications dealing with the problems of this book. Second, during the nine years since the first edition was issued many new facts have come to light which amplify and strengthen the general hypothesis of the effect of climate upon civilization.

The present edition differs from the first in several important respects:

1. In the first edition inheritance, physical environment, and culture were recognized as the three main factors in determining the distribution of civilization. Physical environment was of course treated fully, since it is the main subject of the book. Enough was also said about human culture to show that I fully appreciate its importance, especially as an explanation of the difference between aboriginal America and the Old World. Inheritance, however, was dismissed briefly. In the present edition it receives a good deal of emphasis, especially in the first chapter, which is almost wholly new. It would be emphasized much more strongly were it not that in *The Character of Races* I have devoted a whole book to the problem. That book and this are so closely allied that neither is complete without the other.

2. The relation of climate to health has been much discussed during the past nine years. Accordingly three new chapters, VII

to IX, have been added on this topic. They balance the three preceding chapters by discussing the manifold effects of climate and weather upon man from the standpoint of disease and death, instead of from the standpoint of the day's work.

3. In the first edition of *Civilization and Climate* I assumed that historians and others would be more familiar with the evidence of climatic changes during historic times than is actually the case. Accordingly in Chapter XIV the hypothesis of pulsatory climatic changes is more fully elaborated than formerly, while Chapter XV, which is new, is devoted to criticisms of that hypothesis.

4. Chapters XVI and XVII are devoted to some of the main criticisms of the hypothesis that climate is one of the three main determinants of the distribution of civilization. An especially important new feature is a study of the white man in tropical Australia.

5. In addition to this a large number of minor points have been added. Hence, although certain sections such as Chapters III to VI remain practically unchanged, the book as a whole has a distinctly new aspect. In its present form the book does not insist as strongly as before upon the supreme importance of climate, but the arguments which lead to the conclusion that climate ranks with racial inheritance and cultural development as one of the three main factors in determining the distribution of civilization seem much stronger than previously.

E. H.

New Haven, September, 1924.

AUTHOR'S BIBLIOGRAPHY

THIS list contains the titles of publications of the author which deal with the problems of this book. For details as to many points here discussed, the reader is referred to these publications.

(A) *Books and longer articles on changes of climate and their effect on man:*

(1) EXPLORATIONS IN TURKESTAN. Publications 26 and 73 of the Carnegie Institution of Washington, 1905, pp. 157-317.

(2) THE PULSE OF ASIA. Boston, Houghton Mifflin Company, 1907, 415 pp.

(3) PALESTINE AND ITS TRANSFORMATION. Boston, Houghton Mifflin Company, 1911, 433 pp.

(4) THE CLIMATIC FACTOR AS ILLUSTRATED IN ARID AMERICA. Publication 192 of the Carnegie Institution of Washington, 1914, 341 pp.

(5) THE SOLAR HYPOTHESIS OF CLIMATIC CHANGES. *Bulletin of the Geological Society of America,* vol. 25, 1914, pp. 477-590.

(B) *Shorter articles dealing with phases of the problem of climatic changes not treated under (A):*

(6) THE BURIAL OF OLYMPIA. *Geographical Journal,* London, 1910, pp. 657-686.

(7) THE OASIS OF KHARGA. *Bulletin of the American Geographical Society,* New York, vol. 42, 1910, pp. 641-661.

(8) TREE GROWTH AND CLIMATIC INTERPRETATIONS. In *Postglacial Climatic Changes,* Publication No. 352, Carnegie Institution of Washington, 1924. (In press.)

(C) *Books and articles dealing with climate and civilization:*

(9) PHYSICAL ENVIRONMENT AS A FACTOR IN THE PRESENT

CONDITION OF TURKEY. *Journal of Race Development*, Worcester, Mass., vol. 1, 1911, pp. 460-481.

(10) GEOGRAPHICAL ENVIRONMENT AND JAPANESE CHARACTER. *Journal of Race Development*, Worcester, Mass., vol. 2, 1912, pp. 256-281. (Reprinted in a volume entitled JAPAN AND JAPANESE AMERICAN RELATIONS, edited by G. H. Blakeslee, New York, 1912, pp. 42-67.)

(11) CHANGES OF CLIMATE AND HISTORY. *American Historical Review*, vol. 18, 1913, pp. 213-232.

(12) THE ADAPTABILITY OF THE WHITE MAN TO TROPICAL AMERICA. In *Latin America*. Clark University Addresses, 1913, edited by G. H. Blakeslee, pp. 360-386.

(13) THE GEOGRAPHER AND HISTORY. *The Geographical Journal*, London, January, 1914.

(14) A NEGLECTED FACTOR IN RACE DEVELOPMENT. *The Journal of Race Development*, Worcester, Mass., vol. 6, 1915, pp. 167-184.

(15) CLIMATIC VARIATIONS AND ECONOMIC CYCLES. *The Geographical Review*, vol. 1, March, 1916, pp. 192-201.

(16) MAYA CIVILIZATION AND CLIMATIC CHANGES. *Proceedings of the 19th International Congress of Americanists*, Washington, 1917, pp. 150-164.

(17) CLIMATIC CHANGE AND AGRICULTURAL EXHAUSTION AS ELEMENTS IN THE FALL OF ROME. *The Quarterly Journal of Economics*, vol. 31, February, 1917, pp. 173-208.

(18) THE RED MAN'S CONTINENT. New Haven, 1919.

(19) PRINCIPLES OF HUMAN GEOGRAPHY. New York, 1920.

(20) BUSINESS GEOGRAPHY. New York, 1922.

(D) *Books and articles published since the first edition of* CIVILIZATION AND CLIMATE *and dealing largely with the relation between climate and health:*

(21) WORLD POWER AND EVOLUTION. New Haven, 1919.

(22) THE INTERPRETATION OF THE DEATH RATE BY CLIMOGRAPHS. *Modern Medicine*, vol. 1, May, 1919.

(23) THE CONTROL OF PNEUMONIA AND INFLUENZA BY THE WEATHER. *Ecology*, vol. 1, January, 1920.

(24) THE PURPOSE AND METHODS OF AIR CONTROL IN HOSPITALS. *The Modern Hospital*, vol. 14, April, 1920.

(25) METHODS OF AIR CONTROL AND THEIR RESULTS. *The Modern Hospital*, vol. 14, May, 1920.

(26) THE RELATION OF HEALTH TO RACIAL CAPACITY: THE EXAMPLE OF MEXICO. *The Geographical Review*, vol. 11, April, 1921.

(27) AIR CONTROL AS A MEANS OF REDUCING THE POSTOPERATIVE DEATH RATE. *American Journal of Surgery*, vol. 35, July and October, 1921.

(28) INFLUENZA AND THE WEATHER IN THE UNITED STATES IN 1918. *The Scientific Monthly*, vol. 17, November, 1923.

(29) CAUSES OF GEOGRAPHICAL VARIATIONS IN THE INFLUENZA EPIDEMIC OF 1918 IN THE CITIES OF THE UNITED STATES. *Bulletin of the National Research Council*, vol. 6, July, 1923.

(E) *Books and articles dealing with natural selection and racial character:*

(30) GEOGRAPHY AND NATURAL SELECTION. *Annals of the Association of American Geographers*, 1924.

(31) THE CHARACTER OF RACES. New York, 1924.

(F) *Books dealing with the nature and causes of climatic changes:*

(32) CLIMATIC CHANGES (with S. S. Visher). New Haven, 1922.

(33) EARTH AND SUN: AN HYPOTHESIS OF SUNSPOTS AND WEATHER. New Haven, 1923.

CHAPTER 1

CIVILIZATION AND CLIMATE

INTRODUCTION

THE races of the earth are like trees. Each according to its kind brings forth the fruit known as civilization. As russet apples and pippins may grow from the same trunk, and as peaches may even be grafted on a plum tree, so the culture of allied races may be transferred from one to another. Yet no one expects pears on cherry branches, and it is useless to look for Slavic civilization among the Chinese. Each may borrow from its neighbors, but will put its own stamp upon what it obtains. The nature of a people's culture, like the flavor of a fruit, depends primarily upon racial inheritance, which can be changed only by the slow processes of biological variation and selection.

Yet inheritance is only one of the factors in the development of civilization. Religion, education, government, and all of man's varied occupations, customs, and institutions—his inherited culture as the anthropologists say—form a second great group of social influences whose power seems almost immeasurable. They do for man what cultivation does for an orchard. One tree may bear a few wormy, knotty little apples scarcely fit for the pigs, while another of the same variety is loaded with great red-cheeked fruit of the most toothsome description. The reason for the difference is obvious. One tree grows in a grassy tangle of bushes with no room to develop, little chance to get sunlight, and scant opportunity to obtain nourishment because of the abundance of other plants and the poverty and thinness of the unfertilized soil. The other stands in the midst of a carefully tilled garden where it has plenty of room to expand and enjoy the sun, and where its roots can spread widely in a deep, mellow, well-fertilized soil. Moreover, one tree is burdened with dead wood and suckers, and infested with insects and other parasites; while the other is carefully pruned, scraped, and sprayed.

In spite of the most careful and intelligent cultivation, a tree of the finest variety may fail to produce good fruit. Too much rain or too little; prolonged heat or constant cloudiness; frost when the blossoms are opening, or violent wind and hail may all be disastrous. The choicest tree without water is worth less than the poorest where the temperature and rainfall are propitious. Its health is ruined, and it can bear no fruit. Here, as in the preceding case, the great need of the tree is health in the fullest and broadest sense. A good climate, good cultivation, and good nourishment are merely means of giving the tree perfect health and thus allowing the fullest development of its inheritance. Thus the two great factors which really determine the quality of the fruit are inheritance and health. The other factors, namely, food, climate, parasites, and cultivation, are important chiefly as means whereby health, or, perchance, inheritance, is modified.

Does the fruit known as civilization depend upon these same conditions? It seems to me that it does. Few would question that a race with a superb mental and physical inheritance and endowed with perfect health is capable of adding indefinitely to the cultural inheritance received from its ancestors, and thus may attain the highest civilization. But if that same cultural inheritance were given to a sickly race with a weak inheritance of both mind and body, there would almost surely be degeneration. Aside from biological inheritance, the main factors in determining health are climate, food, parasitic diseases, and a people's stage of culture, which corresponds to the cultivation of the tree. Moreover, these same four factors, through their potency in selecting some types for preservation and others for destruction, and perhaps through their power to cause mutations, are among the main agencies in determining inheritance. Climate stands first, not because it is the most important, but merely because it is the most fundamental. It is fundamental by reason of its vital influence upon the quantity and quality not only of man's food but of most of his other resources; it plays a large part in determining the distribution and virulence of the parasites which cause the majority of diseases; and through its effect upon human occupations, modes of life, and habits, it is one of the main determinants of culture. On the other hand, neither food, disease, nor culture has any appreciable effect upon climate, although they may modify its influence. Moreover, climate has a direct effect upon health in addition to its indirect effect through food, disease, and mode of life. Hence although climate may be no more important than other factors in determining the relative degree of progress in different parts of the world, it is more fundamental in the sense that it is a cause rather than a result of the other factors.

In studying climate it is essential to draw a sharp distinction between three types of influences. In the first place climate has a direct effect upon man's health and activity. Second, it has a strong indirect but immediate effect through food and other resources, through parasites, and through mode of life. Third, by its combined direct and indirect effects in the past it has been a strong factor—some would say, the strongest—in causing migration, racial mixture, and natural selection; and it may have had something to do with producing the variations which the biologists call mutations. Thus it has had a powerful effect upon inheritance.

From the days of Aristotle to those of Montesquieu and Buckle, there have been good thinkers who have believed that the direct effect of climate is the most important factor in determining the differences between the degree of progress in various parts of the earth. Others have held that wherever food is available for a moderately dense population and man can avoid diseases like tropical malaria, human culture can rise to the highest levels. The location of the world's great nations seems to them largely a matter of accident.

The majority of people reject both of these extreme views. Few doubt that climate has an important relation to civilization, but the majority consider it less important than racial inheritance, proper food, or good institutions in the form of church, state, and home. We realize that a dense and progressive population does not live in the far North or in deserts simply because the difficulty of getting a living grinds men down and keeps them isolated. We know that the denizens of the torrid zone are slow and backward, and we almost universally agree that this is connected with the damp, steady heat. We continually give concrete expression to our faith in climate. Not only do we talk about the weather more than about any other one topic, but we visit the seashore or the mountains for a change of air. We go South in winter, and to cool places in summer. We are depressed by a series of cloudy days, and feel exuberant on a clear, bracing morning after a storm. Yet, in spite of this universal recognition of the importance of climate, we rarely assign to it a foremost place as a condition of civilization. We point out that great nations have developed in such widely diverse climates as the hot plains of Mesopotamia and Yucatan and the cool hill country of Norway and Switzerland. Moreover, although Illinois and southern Mongolia lie in the same latitude and have the same mean temperature, they differ enormously in civilization. To put the matter in another way, we recognize two great sets of facts which are apparently contradictory. We are conscious of being stimulated or depressed by climatic conditions, and we know that as one goes northward or southward, the distribution of civilization is generally in harmony with what we should expect on the basis of our own climatic experiences. Nevertheless, even in our own day, regions which lie in the same latitude and apparently have equally stimulating climates differ greatly in their degree of civilization. When we compare the past with the present, we find the same contradiction still more distinctly marked. Hence our confusion. From personal experience we know that the direct effects of climate are of tremendous importance. Yet many facts seem to indicate that this importance is less than our observation would lead us to anticipate.

The reason for this doubtful attitude can easily be discovered. The things that we call facts are often not well established. Although we believe in the influence of climate, we know little of the particular climatic elements which are most stimulating or depressing. How much do we know of the relative importance of barometric pressure, wind, temperature, or humidity? What about the comparative effects of the climates of England and southeastern Russia? In addition to this, we are far from knowing what type of climate prevailed in Egypt, Greece, or Mesopotamia when they rose to eminence. Many good authorities have asserted that the climate of those regions was the same two or three thousand years ago as now. This view is rapidly losing ground, but those who believe in a change are not certain of its nature. They are not yet wholly agreed as to whether it has produced an important influence upon the particular climatic elements which are most stimulating to the human system.

This book has been written because two recent lines of investigation apparently combine to explain at least part of the contradictions which have hitherto proved so puzzling. In the first place a prolonged study of past and present climatic variations led to the conclusion that the climate of the past was different from that of the present. In early historic times, for example, some parts of the world appear to have been drier than now, and other parts moister. In any given place, however, the change from the past to the present has not consisted of a steadily progressive tendency in one direction but of variations. In the places that were formerly moister than now there appear to have been alternate changes toward dryness and then toward moisture; while in the places that are drier than in the past there have been corresponding variations of the opposite types. In a word, the climate of historic times seems to have undergone a pronounced series of pulsations which have varied in character from one part of the earth to another. The second line of investigation which originally led to the writing of this book was a study of the climatic conditions under which people of European races are able to accomplish the most work and have the best health. This investigation led to the conclusion that the principle of climatic optima applies to man quite as fully as to plants and animals. According to this principle each living species has the best health and is most active under certain definite conditions of temperature, humidity, wind movement, storminess, variability, and sunlight, or, more exactly, under certain combinations of these conditions. Any departure from the optimum conditions leads to a decrease of activity and efficiency.

During the last ten years the importance of racial inheritance and racial selection has been strongly emphasized. In the first edition of *Civilization and Climate* the importance of race is strongly emphasized, but I failed to see how important a part has been played by climatic changes in selecting certain types of people for destruction or preservation. Such selection is apparently one of the chief ways in which the character of races is altered. The climatic pulsations of the glacial, post-glacial, and historic periods appear to have exerted a profound influence upon the degree of habitability of different parts of the earth. Thus famine, distress, and disease have arisen, and the pressure of population has led to migration, racial mixture, and the preservation of one type of people in one place and another somewhere else. Natural selection under the stress of climate goes far toward explaining many of the cases where the distribution of civilization does not agree with what would be expected on the basis of the direct effect of climate. So important is this that I have written a book on *The Character of Races* for the express purpose of applying the principles of natural selection to the history of racial development. That book might well have been called *Civilization and Race* in order to emphasize the fact that it is a continuation of the present work. The change in my own realization of the part played by climatic changes is one of the chief reasons why the present edition of *Civilization and Climate* is in some respects almost a new book.

A large part of the reasoning of this book stands or falls with the hypothesis of climatic pulsations in historic times. The steps which led to the hypothesis may be briefly sketched as follows: In 1903, under the inspiration of the broad vision of Raphael Pumpelly and the careful scientific methods of William Morris Davis, I began to study the climate of the past. Two years' work with the Pumpelly Expedition sent to Turkestan by the Carnegie Institution of Washington led to the conviction that Reclus, Kropotkin, and others are correct in believing that two or three thousand years ago the climate of Central Asia was moister than now, a view which I advocated in *Explorations in Turkestan*. Later, during the Barrett Expedition to Chinese Turkestan, it became evident that the scientists who hold that the ancient climate in those regions was as dry as that of today also have much strong evidence to support their view. It soon appeared, however, that this apparent contradiction is fully explained by the fact that throughout the dry regions of Central Asia and the eastern Mediterranean the evidences of moist and dry conditions, respectively, are grouped in distinct periods; the beginning of the Christian era was moist, for example, and the seventh century dry. This led to what I have called the "pulsatory hypothesis," which furnished a name for *The Pulse of Asia*. According to this hypothesis, although the historic and prehistoric past in those particular regions was in general moister than the present, the change from moist to dry has taken place

irregularly in great waves. Even in early historic times certain
centuries were apparently drier than today, while others not
long ago were moist. In 1909, this view was confirmed during
the Yale Expedition to Palestine, the results of which are set
forth in *Palestine and Its Transformation.*

Then a series of journeys in the drier parts of the United
States and in Mexico and Central America, in coöperation with
D. T. MacDougal of the Desert Botanical Laboratory of the
Carnegie Institution, showed that the main features of previous
conclusions apparently apply to the New World as well as the
Old. The most important feature of this work in America was
the measurement of the thickness of the annual rings of growth
of some four hundred and fifty of the Big Trees of California—
the Sequoia Washingtoniana, which grows high in the Sierras.
These measurements made it possible to form a fairly reliable
climatic curve for 2000 years and an approximate curve for
another thousand. The final data as to the Big Trees were published in *The Climatic Factor,* which appeared in 1915, at the
same time as *Civilization and Climate.* The agreement of the
California curve with the climatic curve of western Asia as previously worked out, and the constantly growing evidence as to
the reliability of tree growth as a measure of climate have done
far more than anything else to cause the hypothesis of climatic
pulsations to be widely accepted. Here is the way the matter
is summed up by the British meteorologist, C. E. P. Brooks, in
his book on *The Evolution of Climate* (1922): "The question of
climatic changes during the historic period has been the subject
of much discussion, and several great meteorologists and geographers have endeavoured to prove that at least since about 500
B.C. there has been no appreciable variation. It is admitted that
there have been shiftings of the centers of population and civilization, first from Egypt and Mesopotamia to the Mediterranean regions, and later to northern and western Europe, but
these have been attributed chiefly to political causes and es-

pecially to the rise of Islam and the rule of the 'accursed Turk.' Recently, however, there has arisen a class of evidence which cannot be explained away on political grounds, and which appears to have decided the battle in favour of the supporters of change; I refer to the evidence of the trees. The conclusions derived from the big trees of California have fallen admirably into line with archæological work in Central America, in central Asia and other regions, and have shown that the larger variations even of comparatively recent times have been very extensive, if not world-wide, in their development."

Another important factor in perfecting the pulsatory hypothesis has been the study of the Maya ruins in Yucatan and Guatemala. They join with other evidence in suggesting that changes of climate are of different types in different parts of the world. Central America seems to have been relatively dry at times when the Big Trees of the Sierras suggest that California was moist. This is an important modification of some of the conclusions which I have seemed to imply in earlier books. At practically the same time when this newer conclusion was published, an almost identical idea was presented by J. W. Gregory of England, whose article *Is the Earth Drying Up?* in the *Geographical Journal* for 1914 is the strongest criticism of my climatic theories that has ever appeared. It will be discussed more fully later. The fact that two investigators who seemed to be opposed should independently publish the same conclusion without knowing what the other was doing greatly strengthens the force of that conclusion.

Having reached the conclusion that pulsatory climatic changes have taken place during historic times and have differed in type from region to region, the next step was to study the mechanism and cause of the supposed changes. In Palestine and the eastern Mediterranean the conditions of vegetation, especially the palm and vine, as Gregory has well shown, make it almost certain that variations in storminess and rainfall, rather than in temperature, have been the primary factor. The recorded observations upon the mild climatic pulsations of the past hundred years support this conclusion. Various lines of evidence also indicate that climatic pulsations probably consist of a shifting of the earth's climatic zones or at least of the areas of cyclonic storms alternately toward and away from the equator. The idea as to zones, although not as to storms, was announced by the German geologist, Penck, at essentially the same time that I announced it, but the two conclusions were wholly independent and were based on quite different data. In both cases it was specifically recognized that the same kinds of climatic shiftings have taken place both in prehistoric and historic times, although the earlier changes were of greater magnitude. In these shiftings the temperate zone of storms appears to have been shoved irregularly back and forth. When it was farther south than at present, the subtropical countries which now are subarid must have been relatively moist. At the same time the subtropical arid belt was apparently shifted toward the equator so that on the borders of the torrid zone certain lands which now are wet were then relatively dry. When the shifting of zones took place in the opposite direction the reverse changes of climate apparently took place.

It was only after the preceding conclusions as to climatic pulsations had reached essentially their present form that I began the next phase of the investigation, namely, the study of the actual effect of present climates upon human health and activity. This is important because some critics have supposed that I have unduly emphasized the importance of climatic changes, or have even formed a theory in regard to them for the purpose of bolstering up a preconceived idea that differences of climate from place to place are a main cause of the present distribution of human progress. On the contrary, up to this period my reasoning had been somewhat as follows: If climatic changes have occurred during historic times, they must have had some economic effect because such changes alter the capacity of a region to support population. The economic changes in their turn must have led to political disturbances and migrations. Is there any evidence of such events at times when the climate suffered unusually great or rapid changes?

The possibility of such a connection between climate and history has deeply interested a great number of students. Kropotkin, for instance, has vividly portrayed the way in which a gradual desiccation of Asia presumably drove into Europe the hordes of barbarians whose invasions were so important a feature of the Dark Ages. If the change from the climate of the past to that of the present has been marked by pulsations rather than by a progressive change in only one direction and if there have been certain periods of rather rapid change and of great, though temporary, extremes, as seems highly probable, the correspondence between historic events and climatic vicissitudes may be closer than would otherwise seem credible. Indeed, as soon as I had framed a preliminary outline of the curve of climatic changes during historic times, it appeared as though many of the great nations of antiquity had risen or fallen in harmony with favorable or unfavorable conditions of climate. During periods of drought, not only are the people of the drier regions forced to migrate, especially if they are nomads, but increasing aridity, even in more favored places such as Greece, must cause economic distress, and thus engender famine, misery, and general discontent and lawlessness. A recent journey to China, which gave an opportunity for a study of the famines and barbarian invasions that have afflicted that country for two thousand years, has added greatly to the already abundant evidence of the truth of this view. It has also emphasized the remarkably intimate connection between economic distress and political discontent, a connection which is obvious in advanced countries like the United States, Australia, and Europe as well as in backward regions like China, Persia, India, and Mexico.

While these economic and political effects of climatic changes were being studied I became more and more impressed by the fact that when each country rose to a high level of civilization, it appears in a general way to have enjoyed a climate which approached more closely than now to certain well-defined conditions. These conditions appeared to resemble, but by no means duplicate, those now prevailing in most of the regions where civilization is highest. In spite of marked variations, the general tendency during periods of high civilization has apparently been toward cool, but not extremely cold, winters, and toward summers which though warm or even hot for several months are generally varied by storms or at least by winds which produce frequent changes of temperature. It became especially evident that a relatively high degree of storminess and a relatively long duration of the season of cyclonic storms have apparently been characteristic of the places where civilization has risen to high levels both in the past and at present. Hence such places experience much variability, a condition which later work has led me to believe highly beneficial.

Up to this point in my investigations, I saw no ground for appealing to anything except economic and political factors in explanation of the apparent connection between civilization and climate. Then a little book on *Malaria: A Neglected Factor in the History of Greece and Rome,* by W. H. S. Jones, convinced me that climatic changes have altered the conditions of health as well as the economic situation. Later studies indicate that in other countries such as Central America, Indo-China, Java, and Egypt, as well as Greece and Rome, changes in the amount and virulence of such diseases as malaria and yellow fever may have been potent factors in diminishing the vitality of a nation. In fact it now seems probable that through their effect on bacteria, on the water supply, on the breeding places of insects, on the quality of the food, and perhaps in other ways climatic changes may exert quite as much effect as through the more direct economic channels.

The study of diseases was the natural prelude to a closer inquiry into the fact that at times of favorable climate in countries such as Egypt and Greece the people were apparently filled with a virile energy which they do not now possess. Many authorities attribute the loss of this to an inevitable decay which must overtake a nation as old age overtakes an individual. Others ascribe it to the lack of adaptability in various institutions, to increasing luxury, to contact with inferior civilizations, to a change in racial inheritance, or to various other factors, most of which are doubtless of importance. Previous to 1911 a few authorities such as O. Fraas had connected the decline of energy in Egypt, for example, with climatic changes, but they gave so few reasons and the whole matter seemed so doubtful that I had little faith in their suggestions. At that time, Charles J. Kullmer of Syracuse University sent me a manuscript calling attention to the remarkable similarity between the distribution of cyclonic storms and of civilization. His article was never published, but was presented at a meeting of the Association of American Geographers. He advanced the idea that the barometric changes which are the primary cause of storms, or perhaps some electrical phenomena which accompany them, may produce a stimulus which has much to do with the advancement of civilization. Although he presented no definite proof, his suggestion seemed so important that I determined to carry out a plan which had long been in mind. This was to ascertain the exact effect of different types of climate by means of precise measurements. Dexter, in his book on *Weather Influences*, had made a beginning. Lehmann and Pedersen had made a small series of measurements whose highly suggestive results have been published under the title *Das Wetter und unsere Arbeit*. A few physicians and students of child psychology were also at work, and their results have been summed up in such publications as Hellpach's *Geopsychische Erscheinungen* and Berliner's *Einfluss von Klima, Wetter und Jahreszeit auf das Nerven- und Seelen-leben.* Nevertheless, there existed no large series of measurements of the actual efficiency of ordinary people under different conditions of climate.

The ideal way to determine the effect of climate would be to take a given group of people and measure their activity daily for a long period, first in one climate, and then in another. This, however, would not be practicable because of the great expense, and still more because the results would be open to question. If people were thus moved from place to place, it would be almost impossible to be sure that all conditions except climate remained uniform. If the climate differed markedly in the two places, the houses, food, and clothing would also have to be different. Social conditions would change. New interests would stimulate some people and depress others. Hence no such experiment now seems practicable.

The most available method is apparently to take a group of people who live in a variable climate, and test them at all seasons. The best test is a man's daily work, the thing to which he devotes most of his time and energy. Accordingly, I took the records of four groups of people; namely, some five hundred factory operatives in three Connecticut cities, New Haven, New Britain, and Bridgeport; nearly nine thousand operatives at Pittsburgh; three or four thousand operatives in southern cities from Virginia to Florida; and over seventeen hundred students at the United States Naval Academy at Annapolis, and the Military Academy at West Point. In most cases each person's record covered an entire year, or at least the academic year. All the records were compared with the weather, as explained in later chapters. The results were surprising. Changes in the barometer seemed to have little effect. Humidity apparently possesses considerable importance, but the most important element is clearly temperature. The people here considered were physically most active when the average temperature ranged from 60° to 65° F., that is, when the noon temperature rose to 70° or more, while the night temperature fell to 55° or so. This is higher than many of us would expect. Mental activity, however, reached a maximum when the outside temperature averaged about 38° F., that is, when there were mild frosts at night. Another highly important climatic condition seemed to be the change of temperature from one day to the next. People did not work well when the temperature remained constant. Great changes were also unfavorable. The ideal condition, or optimum, seemed to be mild winters with some frosts, mild summers with the temperature rarely above 75° F., and a constant succession of mild storms and moderate changes of weather from day to day.

The facts just stated seem to be of great significance as will be fully explained in this book. They suggest that the weather exerts a rather large and easily measurable effect upon man's capacity for both physical and mental work. This conclusion naturally led to a query as to how the climates in different parts of the world vary in their effect on human efficiency. Accordingly, I constructed a map showing how human energy would be distributed throughout the world if all the earth's inhabitants were influenced as were the fifteen thousand people of the four American groups mentioned above. This map was found to agree to a remarkable extent with a map of the present distribution of civilization based on the opinions of about fifty geographers and other widely informed men in a dozen countries of America, Europe, and Asia. Moreover, it agreed with the conclusions previously drawn as to the relation of climatic changes to the civilization of the past. Take, for example, the decadent countries as to whose past climate we have some definite idea. In practically every case the climate during their more flourishing periods appears to have approached the optimum, as determined in the United States, more nearly than at present. This does not mean that the climate of Egypt, North India, China,

Greece, or the Maya regions in Guatemala was ever like that of either New York or California. It merely means that it approached more closely than at present to one or the other of these American types. Hence at the time of its greatness each region apparently enjoyed more than its present advantages in economic conditions, in freedom from parasitic diseases, and in direct climatic stimulus.

When my investigations had reached this point, the first edition of *Civilization and Climate* was written. During the nine or ten years that have since elapsed not only has much new evidence come to light, but my own point of view has changed considerably. The changes are set forth in the series of books and articles listed in the preface to the present (third) edition. They are referred to at some length in later chapters, but may here be briefly summarized.

The comments on *Civilization and Climate* by historians and others make it more and more evident that the crux of the hypothesis of this book lies in changes of climate. The question, however, is not whether the climate of ancient Egypt, for example, was like that of modern England. It certainly was not and never could be. The contrast of the two countries in latitude, topography, and relation to land and sea makes any such close resemblance impossible. The real question is this: When Egypt rose to its greatest heights did its climate approach appreciably nearer than now to the type which provides the optimum conditions of energy, health, and economic strength for a people in the early Egyptian stage of development? Bear in mind that when the ancient Egyptians first rose to a state approaching civilization, they had not yet learned to use iron tools. Even in later days they had nothing like our modern skill in using wood and coal for heating houses, in manufacturing cotton and woolen cloth on such a scale that even the poor can be warmly clothed, in building houses that are proof against wind, rain, and cold; nor had they our skill in combating disease. Hence their stage of development caused the optimum climate to be warmer for them than for us. We are able to guard ourselves against low temperature and exposure, and thus gain an important stimulus without suffering much harm. They could not withstand cold winters.

Bearing in mind then, that the optimum climate varies according to a nation's stage of civilization and also that there is doubtless some difference in the optimum from race to race, our problem becomes to determine how far the climate of the past in any given region was like that which is best for the stage of human progress, and perhaps the race, with which we happen to be dealing. This point of view is slightly different from that of the first edition of this book. The change is due largely to GilFillan's article on *The Coldward Course of Progress* in the *Political Science Quarterly*, 1920.

So far as changes of climate are concerned, the ten years since this book was written have seen considerable new evidence as to the reality and nature of historic as well as prehistoric pulsations. As an entire new chapter will be devoted to this matter, and as the comment of Brooks on the convincing character of the evidence has already been quoted, it is enough to point out here an interesting fact as to the kind of people who have accepted the conclusions of this book. The non-scientific public, which has doubtless been the widest audience of *Civilization and Climate*, has accepted the book on the reasonableness of its main hypothesis, and with an open mind as to future proof or disproof. Geologists, archæologists, and those geographers who have had a geological training are the types of scientists who have found the hypothesis most convincing. This is because most of the methods of reasoning and lines of evidence employed in discussions of climatic changes are such as are commonly used in geology, archæology, and geography. The evidence consists chiefly of ancient lakes and streams, old roads in deserts, dead vegetation and its rings of annual growth, abandoned fields and irrigation systems, and especially ruins and other traces of man, which are really human fossils. Such evidence appeals to geologists, archæologists, and geologically trained geographers.

Anthropologists, economists, and historians, on the other hand, have been slow to believe that climatic changes have had much influence upon human history. They accept, indeed, the geologists' conclusion that previous to recorded history great climatic changes drove man this way and that, destroyed ancient types of culture, and either wiped whole races out of existence, or profoundly modified them, physically, mentally, and socially. They apparently have no difficulty in accepting the geological evidence that among primitive men, as among plants and animals, climate has been one of the most powerful factors in determining the distribution and vigor of different types. But when it comes to the period of written history many historians, and some anthropologists and economists, no longer trust the geological methods of reasoning. Their opinions are more or less unconsciously molded by two widely accepted assumptions.

The first assumption is that climatic uniformity is a normal condition. This idea seems wholly untenable in view of the constantly growing evidence of numerous and important glacial periods and other extreme types of climate at all stages in geological history. The more we know of the geological record, the more clear it becomes that change, not uniformity, is the rule. Even in the long periods when the larger types of climatic changes have been absent, there is abundant evidence of minor fluctuations and pulsations. The second assumption is equally doubtful. It holds that written records and statistics are more reliable than the geological type of evidence. Of course written records and statistics are far more reliable than any other types of evidence if they are sufficiently full, if they can be trusted, and if they are prepared by people who are conscious of the purposes for which they are to be used. But the written and recorded evidence as to the climate of the past consists of mere

scraps of information set down in most cases accidentally and
with no idea of their possible significance in the distant future.
Such evidence has, of course, great value, and must be studied
assiduously. Nevertheless, it is inevitably subordinate to the
geological type of evidence, which may be either physiographic,
botanical, or archæological.

It seems clear, then, that the ultimate decision as to whether
climatic changes have taken place on a large scale during historical times does not rest with historians. It rests primarily
with persons who are trained in climatological, and especially
geological, methods. During the last ten years geographers as
a whole, in spite of some exceptions, seem to have become per-
suaded that the historical period has witnessed a series of climatic pulsations like those of the prehistoric or post-glacial
period although on a smaller scale. According to their own
written statement in answer to a questionnaire, over nine tenths
of the geographers of America, if we may judge from the fifty
who have expressed their opinions, hold this view, although they
are not quite so fully agreed as to how great the effect of these
pulsations upon man may have been. Of course it is possible
that a few geographers failed to answer the questionnaire because they did not wish to express an opinion contrary to mine.
I think, however, that the number of such is very limited, for
most of those who are known to hold views unlike my own expressed them freely. Even when all due allowance is made for
failures to answer, it seems clear that among the people who
are best competent to weigh the evidence the great majority
believe in pulsations of climate. If these geographers with a
geological training are right, there seems to be no escape from
the conclusion that during certain periods of ancient history the
climate of places like Egypt, Mesopotamia, and North India
approached the optimum more closely than at present. Since the
lower stage of culture in those early days presumably caused
the optimum temperature for human progress to be higher than

is now the case among the most advanced races, the climatic conditions were even more favorable than I realized when this book was first written. If this conclusion is well grounded, it becomes a basic fact which the historian must take into account just as every careful student of early man now continually takes account of the fact that primitive man was greatly influenced by the vicissitudes of the successive glacial epochs, and just as every economist recognizes that modern man, on a smaller scale, is profoundly influenced by good or bad crops. History can never be written correctly until its physical basis is thoroughly understood, and until it is recognized that economic conditions, and human health and energy vary from age to age almost as much as do the conditions of politics, religion, and personality.

Another line where there has been much progress in the last ten years is the determination of the nature and importance of the climatic optimum for man's physical development and health as opposed to his economic development. After *Civilization and Climate* had been completed, I undertook a study of the relation of deaths to climate. The results appeared in *World Power and Evolution.* Some eight and a half million deaths in about sixty cities in the United States and thirty in France and Italy were analyzed according to the average weather of each month during periods which in most cases amounted to at least ten years. Except for about 400,000 in the United States, all the deaths occurred during the normal period immediately preceding the Great War. In addition to this, about 50,000,000 deaths in other countries were analyzed less intensively, but with essentially the same results. On the other hand, about 700,000 deaths in New York from 1877 to 1888 have been very minutely analyzed according to the day of death. For the six years from 1883 to 1888 the number of deaths each day has been compared with the weather day by day during the two weeks ending with the day of death. In another investigation 7200 deaths from lobar pneumonia in New York in 1913 were compared with the weather for each day. Again, in the two largest hospitals in Boston the relation of the weather to 2300 deaths succeeding operations was looked into. At a later date the death rate during the influenza epidemic of 1918 in the United States was analyzed by still a different method in order to determine whether the weather had any effect in altering the rate from city to city.

The net results of these investigations, as shown in *World Power and Evolution* and in various technical articles agree closely with those of the investigation of factory workers and students. They confirm the work of other investigators in showing almost beyond question that there is a distinct optimum condition of climate for man just as for plants and animals. This optimum varies relatively little from one set of people to another or from place to place. Even for negroes the departure from the white standard is by no means so large as would be expected, though it is unmistakable. Any departure from the optimum for a given race or individual seems to render people not only less efficient, but also more susceptible to disease and hence an easier prey to bacteria and other parasites. Moreover, as appears in *World Power and Evolution*, there is some evidence that departures from the optimum climate render people less buoyant, less capable of prolonged and steady mental activity, and correspondingly less likely to make progress. The significant fact about the whole matter is that, so far as I am aware, in every case where large bodies of people have been carefully analyzed, the same major responses to climate become evident, even though there may be differences in details. Thus the progress of the last ten years seems to add appreciably to the general certainty as to the nature and importance of climatic optima, and as to the effect of departures therefrom upon health, efficiency, and progress.

The difference between the present and first editions of this book in respect to natural selection and racial inheritance is especially important. It may be illustrated by ancient Greece.

In the first edition I supposed that the reason for the peculiar ability of Greece was a mystery which there was no immediate prospect of solving. It was the result of some unexplained biological process. I believed that no matter where those particular people might have gone or at what period they had happened to live they would have achieved much more than any ordinary people. As events actually shaped themselves, the ancient Greeks migrated to a land near more ancient centers of civilization whence they could receive the inspiration of the greatest preceding cultures. By reason of the numerous gulfs and harbors of their land the Greeks were easily able to get what they wanted from other countries, provided they had energy enough to sail abroad for material riches, and capacity enough to absorb the mental riches with which they came in contact. So far as climate is concerned, Greece appears to have enjoyed unusually favorable conditions throughout most of the period from perhaps 1000 to 300 B.C., and especially about 400 B.C. Previous to 1100 B.C., however, there seems to have been an unfavorable period culminating perhaps 1300 to 1200 years before Christ, while at a later date a period of rapid climatic degeneration from 300 to 200 B.C. was followed by highly unfavorable conditions during the succeeding century. The favorable climate during the period of the greatest Grecian development apparently rendered the economic conditions distinctly more favorable than those of today, and stimulated the Greeks to a high degree of physical and mental energy. At the same time it rendered the environment unfavorable not only to the anopheles mosquito which causes malaria, but to other disease-bearing organisms. Thus the Greeks with their high inheritance enjoyed an environment which gave full opportunity for the development of the best that was in them. The fall of Greece, according to the hypothesis set forth in this book ten years ago, was greatly influenced by a rapid deterioration of climate. In the third century B.C., a decrease in rainfall caused a most serious diminution in the capacity of Greece to support population. This increased the political difficulties while at the same time the ability of the people to cope with such difficulties was diminished by a decline in the stimulating qualities of the climate. At the same time the increase in marshes and in stagnant pools, because of changed conditions of rainfall, made the environment favorable for malarial mosquitoes, while poverty, a poor food supply, and other adverse conditions also fostered disease. In general, the poor economic and political situation, and the unfavorable conditions of health tended to kill off the blonde Norse invaders who seem to have been the leaders among the early Greeks.

At present my view of the rise and fall of Greece differs from the one set forth above in only one respect, but that is highly important. Instead of recognizing natural selection as an important cause merely of the decline of Greece, I now regard it as one of the main causes of the rise of that country. The people who made Greece great, as I have shown in detail in *The Character of Races*, were primarily the Dorians and especially the Ionians. These people appear to have originated far to the north, perhaps in the great plains of Russia. They came to Greece in a series of migrations, the first of which may have been that of the Achæans in the fourteenth century before Christ, while a later invasion culminated in the Dorian invasion two or three centuries later. The Ionians appear to have been largely Greeks of the old Achæan and Minoan upper classes, who were led to migrate by the prolonged disturbances and fighting which harassed Greece for generations and perhaps centuries after the Dorian invasion. Attica had previously been almost depopulated and did not attract the Dorians because it was infertile and poverty-stricken by reason of its dryness. Later, however, perhaps in part because of an amelioration of climate, it became the refuge of large numbers of the old upper classes from the rest of Greece. When its population became too dense, some of the settlers, together with many of the more able and energetic people from other parts of Greece, went across the Ægean Sea eastward to the Ionian coast of Asia Minor. Now we are expressly told by Thucydides and others not only that the settlers in Attica were a highly selected group from the leading families of the rest of Greece, but that they strenuously kept themselves apart from the lower classes. Even if an Athenian married a woman of the lower classes his children followed the social status of the mother and not of the father. Thus among the citizens of Athens a selected inheritance was kept almost unimpaired for many centuries.

But note another important fact. Not only was there a rigid selection of good material when the Ionians settled in Athens, but there appears to have been a perhaps more rigorous selection during the earlier migrations of the people who finally became Greeks. Every migration is more or less selective because the weak, the feeble, the cowardly, and those lacking the spirit of adventure, together with those who lack determination, are gradually weeded out. The longer and more difficult the migration, the more strenuous is the selection, especially among women and children. All but the most vigorous, both mentally and physically, are weeded out with peculiar rapidity. Thus the net result of practically every long or difficult migration in which women as well as men take part is to pick out a relatively small group of people of unusual capacity. The group may contain only one out of twenty, or one out of one hundred or a thousand of the original stock. Now the Athenians, as we have just seen, went through this selection twice, and may have gone through it at earlier times also. They also maintained the quality of the original stock more or less completely by refraining from intermarriage with any but the *élite* of other regions and by various practices such as infanticide, which eliminated weaklings. Such selection with various modifications has been, I believe, one of the most effective factors in producing competent races.

But what has this to do with climate? Much, because a large

number of the migrations of history appear to have been more or less directly started by climatic vicissitudes. For example, the Irish migration to America during the nineteenth century was due largely to the fact that a series of good potato harvests in the early decades of the century permitted the population of Ireland to increase enormously. Then came a series of unduly rainy years accompanied by bad crops. Famine and distress ensued, and immediately an enormous migration to America set in. In later years, as is shown by Brückner's data which I have cited in an article on *A Neglected Factor in Race Development* (*Journal of Race Development*, October, 1915), the migrations from Ireland and likewise from Germany to the United States have varied in close harmony with the climatic conditions and consequent agricultural prosperity of both the Old World and the New. When the crops are bad in Europe but good in America, as is not uncommon, emigration from the Old World is greatly stimulated for a few years. On the contrary a period of bad crops in America coinciding with good crops in Europe gradually diminishes the pressure which leads to migration. Similar examples on a large scale have occurred time after time in the history of China, Central Asia, and other regions.

Contrary to the common belief, most parts of the world normally contain practically as large a population as they are capable of holding under the social and economic conditions which happen to prevail at any given time and place. The birth rate among mankind, as Carr-Saunders convincingly shows in *The Population Problem,* is so large and responds so quickly to economic changes that only a few decades or generations are required for even a relatively depopulated country to fill up to what Woodruff in his *Expansion of Races* has called the saturation point. Thus the great majority of migrations can be understood only by thinking of all parts of the world as normally having nearly their full quota of population. Among primitive hunting tribes, for example, this quota may mean only one person per square mile, while in a modern highly civilized community it may mean a thousand per square mile. The conditions are like those of a bucket filled to the brim with water. As soon as any object is dropped into the bucket some water runs over. In the same way, in most parts of the world any deterioration of economic conditions such as frequently arises from bad crops, immediately necessitates either a lowering of the standard of living or a diminution of population through an increased death rate by reason of poverty, famine, or war, or through migration. A migration, as a rule, is merely a slow drifting of people with unusual energy and initiative from unfavorable to favorable districts. But not infrequently, and especially after one of the sudden climatic vicissitudes which have been common in history, it becomes a violent stream of invasion such as many of the barbarian outpourings about 1400 to 1200 years before Christ and again at various later periods.

The original Greeks apparently took part in extensive wanderings of both the mild and violent kind, and natural selection presumably worked among them effectively. The net result was a highly selected race which gave the world a marvelous group of famous men. These men, it appears, accomplished great things, not only because of their high innate qualities, but because growingly favorable climatic conditions from six to four hundred years before Christ improved their economic situation and helped to give them great energy and splendid health. According to this view, Greece apparently lost her ability partly because the fine original stock became mingled with weaker elements, and partly because the change of climate which began soon after 400 B.C. and which took place most rapidly from 300 to 200 B.C., not only enabled malaria to spread with baleful rapidity, but introduced a stage of diminishing resources, scanty food supply, overpopulation, and lessened climatic stimulus. These factors presumably combined with other well-known conditions to produce poor health and perhaps restriction of families among the upper classes, thus killing off the old dominant stock of northern origin. Hence the racial elements to which Greece owed her greatness disappeared, and the country fell into intellectual insignificance. Other factors undoubtedly coöperated, but such conditions as political corruption, social degeneracy, and undue personal ambition are probably results of racial decay and poor health quite as much as causes. The essential point is that at the beginning a time of climatic stress among nomads in relatively cool grasslands seems to have led to migration and a favorable type of selection. Another period of similar stress amid a highly developed people in a relatively southern land where there was less opportunity or incentive for migration led to the dying off of the more able people. Yet even in this case many of the most competent Greeks migrated, thus still farther impoverishing the mother country and giving rise to highly gifted colonies like the Alexandrian community in Egypt.

To sum up the whole hypothesis of the relation of climate to civilization, here are the factors as I see them at present. Most parts of the world are so well populated that any adverse economic change tends to cause distress, disease, and a high death rate; migration ensues among the more energetic and adventurous people. Perhaps the commonest cause of economic distress is variations in weather or climate which lead to bad crops or to dearth of grass and water for animals. Such economic distress almost inevitably leads to political disturbances and this again is a potent cause of migrations. The people who migrate perforce expose themselves to hardships and their numbers diminish until only a selected group of unusually high quality remains. Such people, either as warlike invaders or in small bands, enter a new country. They may find it well populated and merely impose themselves as a new ruling class, as seems to have happened several times in India, or they may find it depleted of people as in Attica. When the period of climatic stress is ended and the climate improves, the dominant newcomers not only possess an unusually strong inheritance, but are stimulated by unusually good economic conditions and by improved conditions of health and energy. Moreover since the population is apt to remain below the saturation point so long as the climate improves, the standards of living tend to rise and to become relatively high. Thus many people are freed from the mere necessity of making a living and have the opportunity to devote themselves to the development of new ideas in literature, art, science, politics, and other lines of progress. The repeated coincidence between periods of improving climate and periods of cultural progress appears to be due not only to the direct stimulus of climate, as I supposed in the first edition of this book, but to that stimulus combined with a high racial inheritance due to natural selection. This, I am well aware, by no means offers a complete explanation of history, for many other elements must also be considered. But it helps to explain many historic events which have hitherto been only partially understood.

Here, once more, is the sequence—climatic changes produce economic results through an increase or a diminution of the food supply. Thus there arises either a temporary condition of underpopulation with comparative political tranquillity and opportunities for the growth and expansion of civilization, or a condition of overpopulation with consequent political turmoil, war, migration, and the repression of civilization. This point of view was dominant when I wrote *The Pulse of Asia* and *Palestine and Its Transformation*. Climatic changes also appear to have a direct effect in stimulating or repressing man's physical activity, a viewpoint which dominated the first edition of the present book. It is obvious that through their effect upon food, insects, bacteria, and man's own powers of resistance climatic changes must exert a great influence upon disease. Hence I was led to write *World Power and Evolution* from the standpoint of disease and the death rate. But this does not end the matter, for climate apparently exerts a direct selective effect in preserving certain types of people and destroying others, and it certainly exerts these effects indirectly through various conditions already mentioned. Therefore it was only logical that *The Character of Races* should center around natural selection, especially in its climatic aspects. The next step is obviously a study of the relation of climate to mutations and thus to the origin of the new types among which natural selection makes its choice, but that is as yet impossible. Beyond this lies the synthesis of the effects which climate produces through economic and political conditions, through war and migration, food and natural resources, energy and health, and through natural selection and mutations. And all these results of the climatic environment must be put into due relation not only with the results of the other factors of physical environment, but with the opposite side of the shield, that is, with the purely human factors such as institutions, customs, ideas, and all man's passions, ideals, and aspirations. Then it will be possible to form a true philosophy of history. Meanwhile the present edition of *Civilization and Climate* tries to take a broader and deeper view of human progress than its predecessor, but it makes no claim to deal exhaustively with more than one small phase of the matter, namely, the direct effects of climate upon human health and energy.

CHAPTER II

RACE OR PLACE

THE problem which confronts us is primarily to separate the direct effects of climate from those of inheritance, regardless of whether the inheritance has been influenced by the climate of the past. It may be made concrete by comparing two sharply contrasted races, Teutons and negroes. Suppose that there were two uninhabited Egypts, exactly alike, and that one could be filled with negroes and the other with Teutons. Suppose that these settlers were average members of their races, and were equipped with the same religion, education, government, social institutions, and inventions. This might easily happen if the negroes came from the United States. Suppose, further, that neither race received new settlers from without, or lost any except through natural selection. Which would succeed best? "The Teutons, of course," is the answer. "What a foolish question." But is it so foolish? You are thinking of the first generations. I am thinking of the twentieth or later. Does anyone know what five hundred or a thousand years of life in Egypt would do for either Teutons or negroes if no new blood were introduced?

At the end of that time the two sets of people would assuredly be different, for the effect of a diverse inheritance would last indefinitely. The advantage in this respect would presumably be on the side of the Teutons. I wish to emphasize this matter, for I shall have much to say about the effect of climate, and I want to make it perfectly clear that I do not underrate the importance of race. Although the matter is by no means settled, many

authorities think that the brain of the white man is more complex than that of his black brother. Strong, in the *Pedagogical
Seminary* for 1913, and Morse, in the *Popular Science Monthly*
for 1914, have shown that in Columbia, South Carolina, the
white children are mentally more advanced than the colored. By
applying the Binet tests to 225 children in two white schools and
to 125 children in a colored school, they obtained the following
table, showing the amount by which the two races exceeded or
fell short of what would be expected.

	COLORED	WHITE
More than one year backward,	29.4%	10.2%
Satisfactory,	69.8%	84.4%
More than one year advanced,	0.8%	5.3%

Among the white children those from the middle classes made
a better showing than those of factory operatives, but both
were ahead of the colored. So far as home environment is concerned, the factory children have almost no advantage over the
colored children. A slight advantage may possibly arise from
the fact that when the Binet tests were originally devised, they
were designed to measure the capacities of white children. The
negro race may have capacities which the white does not possess
and which do not play a part in the tests. In appreciation of
humor, for example, and in equability of temperament there
can be little question that the black man surpasses the white.
These things, however, can scarcely account for the fact that
29.4 per cent of the colored children showed a mental development more than a year behind that which would be expected from
their age, while only 10.2 per cent of the white children were
equally backward.

So far as I am aware, every exact test which has been made
on a large scale indicates mental superiority on the part of the
white race, even when the two races have equal opportunities.
For example, in Washington the colored children remain in

school quite as long as the white, but they do not accomplish so much in the way of study and do not reach so high a grade. In the cities of the South, Mayo and Loram find that where the races are given essentially the same instruction, the proportion of whites who are promoted is greater than that of negroes. Moreover, the difference seems to increase with years, which suggests that the average colored child not only stands below the average white child in mental development at all ages, but ceases to develop at an earlier age. In the high schools of New York, the superiority of the white race is shown by Mayo's examination of the average marks. By the time the children reach the high school, the processes of promotion have weeded out a much larger proportion of colored children than of white. Hence, the negroes form a specially selected group whose superiority to the average of their race is more marked than the superiority of the white high school children when compared with the rest of the white race. Nevertheless, the average marks of the white children are distinctly higher than those of the colored.

In order to test the capacity of the two races in a wholly different way, I have made a comparison of white and colored workmen employed under precisely similar conditions. The first case was a cigar factory at Jacksonville, Florida. The employees were practically all Cubans. Both the whites and the blacks have very little education, and their home environment in Cuba differs to only the smallest extent. They earn good wages, but are often out of work, and are generally shiftless and unreliable. There is, of course, no color line in Cuba, and the same is true in the cigar factories. Black men and white work side by side at the same tables. In such a factory, if the black man is as capable as the white he has exactly as good a chance, for he is paid by the piece, and his earnings depend entirely on himself. What, then, do we find? Taking all the operatives, we have 39 white and 65 negroes. Their average earnings, as measured by the wages of two weeks, are in the ratio of 100 for the whites to only 51 for the negroes. To make the comparison more favorable to the negroes, let us eliminate those who roll low-grade cigars where little skill is required and the pay is low. We then have 39 white men and 44 negroes. They are doing exactly the same work under exactly the same conditions, but the whites earn a dollar where the negroes earn 75 cents. At a similar factory at Tampa, Florida, 17 colored men were at work and 303 white. In this case practically all of the few negroes happened to be men of long experience, while many of the whites were comparatively new. Nevertheless, the whites are still on a par with the colored men, the ratio being 100 to 99.8.

One of the best places for comparing the two races is the Bahama Islands. For reasons which I shall present later, the process of making "poor whites" has probably gone farther in the Bahamas than in almost any other Anglo-Saxon community. Part of the white people are like their race in other regions, but a large portion have unmistakably degenerated. Witness their intense and bigoted speech, their sunken cheeks and eyes, their sallow complexion, and their inert way of working. In spite of racial prejudice, there is no real color line in the Bahamas. Persons with more or less negro blood are worthy occupants of the highest positions, and are universally accepted in the most exclusive social circles. The British government gives the negro every possible opportunity. The state of affairs may be judged from the remarks of a "poor white" sailor, who said to me: "You want to know why I likes the southern states better than the North. It's because they hates a nigger and I hates him, too. What kind of a place is this where they do everything for the nigger and nothing for the white man? It's bad enough to have to go to jail, but it's damned hard for a white man to be taken there by a nigger constable." In one Bahaman village I saw negro girls teaching white children in the public schools. In that same village a number of the leading white men cannot read or write. When they were children their parents would not send them to school with negroes. The despised negroes learned to read and write, but have now largely forgotten those accomplishments. The proud whites grew up in abject ignorance. Today the same thing is going on. I visited two villages where the white children are staying away from school because they will not go to negro teachers. The homes of such whites are scarcely better than those of their colored neighbors, and their fathers are called "Jim" and "Jack" by the black men with whom they work. Racial prejudice apparently works more harm to the whites than the blacks. So far as occupations go there is no difference, for all alike till the soil, sail boats, and gather sponges.

When the lumber industry was introduced into the islands, whites and blacks were equally ignorant of the various kinds of work involved in cutting trees and converting them into lumber. The managers did not care who did the work so long as it was done. They wanted three things: strength, docility or faithfulness, and brains. They soon found that in the first two the negroes were superior. Time and again persons in authority, chiefly Americans, but also some of the more capable native whites, told me that if they wanted a crew of men to load a boat or some such thing, they would prefer negroes every time. The poor white shirks more than the colored man, he is not so strong, and he is proud and touchy. Other things being equal, the negro receives the preference. But other things are not equal. The very men who praise the negroes generally added: "But you can't use a negro for everything. They can't seem to learn some things, and they don't know how to boss a job." The payroll reflects this. Even though the negroes receive the preference, the 400 who are employed earn on an average only about 60 per cent as much as the 57 white men. If we take only the 57 most competent negroes, their average daily wages are still only 88 per cent as great as those of the native whites. The difference is purely a matter of brains. Although the white man may be ignorant and inefficient, with no more training than the negro, and although his father and grandfather were scarcely better, he possesses an inheritance of mental quickness and initiative which comes into evidence at the first opportunity.

All these considerations seem to point to an ineradicable racial difference in mentality. As the plum differs from the apple not only in outward form and color, but in inward flavor, so the negro seems to differ from the white man not only in feature and complexion, but in the workings of the mind. No amount of training can eradicate the difference. Cultivation may give us superb plums, but they will never take the place of apples. We have tried to convert the black man into an inferior white man, but it cannot be done. Initiative, inventiveness, versatility, and the power of leadership are the qualities which give flavor to the Teutonic race. Good humor, patience, loyalty, and the power of self-sacrifice give flavor to the negro. With proper training he can accomplish wonders. No one can go to a place like Hampton Institute without feeling that there is almost no limit to what may be achieved by cultivation. In an orderly, quiet way, those negro boys and girls go about their daily tasks and give one the feeling that they are making a real contribution to the world's welfare. To be sure, they work slowly, they are not brilliant in their classes, they rarely have new ideas in their manual work, but yet they are faithful. The willing, happy spirit of their work is something that we nervous, worried white people need sorely to learn. Once in a long time there comes a leader, a man to whom both white and black look up, but such leadership is scarcely the genius of the race. Yet leadership is what the black man must have. At such places as Hampton he gets it, and one realizes that the white man's initiative joined to the Christian spirit which is there so dominant can give a training which overcomes much of the handicap of race.

Having turned aside to pay tribute to the potency of race, education, and religion in determining the status of civilization, let us come back to physical environment. What part does this play? Is it so important that a strong race in an unfavorable climate is likely to make no better showing than a weak race in a favorable climate? How far can a bad climate undo the effect of a good training?

In answer to these questions, we may well compare the Teutonic and negro races when each is removed from the climate in which it originally developed. Before proceeding to this a word should be added to forestall any possible misunderstanding of my attitude toward the southern parts of the United States and toward other progressive regions which, nevertheless, suffer somewhat from climatic handicaps. In searching for the truth I shall be forced to say some things which may not be wholly pleasing to residents of such regions. It must be clearly understood, however, that these are not stated on my own authority. All are based either on the consensus of opinion among a large number of persons including many southerners, or upon the exact figures of the United States census or other equally reliable sources. My part has been simply to interpret them. Believing that the South contains a great number of people who in all essential respects have an inheritance equal to that of the best northern stocks, I have tried to find out why the southern part of the United States has prospered less than the northern. This does not mean that I reject the old ideas as to the cause, but simply that I emphasize another which has not received sufficient consideration. It does not discredit the South nor its people. It does not alter the fact that southerners possess a courtesy and thoughtfulness which we of the worried and hurried North need greatly to imitate. Nor does it mean that men of genius are not as likely to be born in one section as another. Instead of this, it merely indicates that in addition to the many efforts now being made to foster progress in the South by other means, we should add a most vigorous attempt to discover ways of overcoming the handicap of climate. This book is written with

the profound hope that the truth which it endeavors to discover
may especially help those parts of the world whose climate, although favorable, does not afford the high degree of stimulation
which in certain other restricted areas is so helpful.

Let us first undertake a study of what the census shows as to
negroes and whites in different parts of the United States. The
only people whom we can compare with accuracy are the farmers,
for they are the only ones for whom exact statistics are available. Fortunately they are the part of the community where
social prejudices and other hampering conditions have the
smallest influence. The prosperity of the farmer, more than that
of almost any other class of society, depends upon his own individual effort. If he is industrious, he need never fear that he and
his family will not have a roof over their heads and something
to eat. Even when the crops are bad, he rarely is in danger of
suffering as factory operatives often suffer, at least not in the
eastern United States, with which alone we are now concerned.
Moreover, the prejudice against colored people has little effect
upon farmers. No one hesitates to buy vegetables peddled by a
darkey farmer. Finally, farming is the occupation in which the
South has been least hampered as compared with the North. For
over half a century the negro has been able to buy land freely in
any part of the country. The southerners, whether white or
black, have suffered economically because of slavery and the consequent war, but they have a good soil and a climate far better
for agriculture than that of the North, and they have peculiarly
good opportunities to raise tobacco and cotton, two of the greatest money-making crops in the world. Taken all in all, the
farmers of the country ought to show the relative capacities of
different races and of the same race under different conditions
better than almost any other class of people.

In 1904 the United States Census Bureau published a bulletin on the negro. From that I have prepared Table 1 showing
the relative conditions in four groups of states in 1900. The first

TABLE 1.*

Comparison of White and Negro Farmers in the Northern and Southern Parts of the United States

N. W. = Northern White Farmers   N. N. = Northern Negro Farmers
S. W. = Southern White Farmers   S. N. = Southern Negro Farmers

	Comparison A				Comparison B			
	New York, New Jersey and Pennsylvania		North and South Carolina, Georgia and Florida		Ohio, Indiana, Illinois, Michigan and Wisconsin		Kentucky, Tennessee, Alabama and Mississippi	
	N. W.	N. N.	S. W.	S. N.	N. W.	N. N.	S. W.	S. N.
1. Total number of farms in 1900. Total	483,772	1,497	406,880	235,720	1,129,810	5,179	635,418	267,530
2. Average acreage per farm. Total	92.5	47.7	133.4	54.6	102.7	55.0	108.0	47.1
3. Average acreage of improved land per farm, and percentage of total land which is improved	63.5 69%	33.0 59%	45.9 34%	31.6 58%	76.5 75%	42.8 77%	50.4 48%	30.6 65%
Average Value:								
4. Farm property per farm. Total	$4,767 100%	$2,801 59%	$1,325 28%	$541 11%	$5,020 100%	$2,227 44%	$1,613 32%	$639 13%
5. Farm property per farm. Land and improvements (except buildings)	$2,516 100%	$1,566 62%	$796 32%	$362 14%	$3,508 100%	$1,646 47%	$944 27%	$405 12%
6. Farm property per farm. Buildings	$1,504 100%	$848 56%	$283 19%	$79 5%	$830 100%	$284 34%	$319 38%	$86 10%
7. Farm property per farm. Implements and machinery	$240 100%	$124 52%	$56 23%	19 8%	$148 100%	$63 43%	$66 45%	$26 18%
8. Farm property per farm. Live stock	$507 100%	$263 52%	$190 37%	$81 16%	$534 100%	$234 44%	$284 53%	$122 23%
9. Products per farm in 1899	$1,025 100%	$516 50%	$501 49%	$318 31%	$981 100%	$473 48%	$515 52%	$360 37%
10. Expenditures per farm for labor in 1899	$104 100%	$48 46%	$42 40%	$13 12%	$60 100%	$23 38%	$26 52%	$11 18%

row of numbers (line 1) shows the total number of white and colored farmers. The second row shows that the farms of the northern white men average about 100 acres in size, while those of the southern white men are larger. The colored farms, on the other hand, have an average size of about 50 acres. In the next row of figures, line 3, we notice that the northerners forge ahead. Even in the relatively hilly states of New York and Pennsylvania, the white farmers have improved 63.5 acres per farm, or 69 per cent of the whole, leaving only 31 per cent in the rough state of bushes or woods. The northern negroes do exactly as well in proportion to their holdings, for they have cleared 33.0 acres, which is also 69 per cent of the average farm. In the Carolinas, Georgia, and Florida, on the contrary, the white men have improved only 34 per cent of their land, and the colored men 58 per cent. For the states farther west (comparison B), approximately similar conditions prevail. The negroes are obliged to clear a larger percentage than the whites because their small holdings would not otherwise furnish a living. The significance of the figures lies in the fact that the northerners, whether white or black, show more energy in improving their land than do the southerners of the same kind.

Since this table was in print the corresponding data for 1910 and 1920 have appeared. Unfortunately they are less full than for 1900 and do not include the value of products, line 9. For line 4 the percentages for each of the three census years are as follows:

TABLE 2.

| | Comparison A | | | | Comparison B | | | |
	N.W.	N.N.	S.W.	S.N.	N.W.	N.N.	S.W.	S.N.
1900	100	59	28	11	100	49	32	13
1910	100	81	41	24	100	48	25	14
1920	100	74	64	20	100	41	31	16

In comparison A, the gain of the other groups in relation to the northern white farmers is noticeable. This, however, does not mean merely that agriculture is improving in the South, but also that it is declining in the Middle Atlantic States. In B, the percentages are almost unchanged.

In both comparisons of Table 1 items 4 and 9 are the most significant. They show the value of the farms and the value of the annual products. In each item of Table 1 the values are stated in dollars as given in the census report, while underneath I have added percentages. In computing the percentages the highest value is reckoned as 100 per cent and the rest figured accordingly. In each item of both comparisons for all three census years, as given in Table 2, the northern whites stand at the top. In general, taking both comparisons into account, the northern white man's farm is worth twice as much as that of his colored neighbor, and he gets twice as much from it. The southern white man has a farm worth less than that of the northern negro, but he gets from it approximately the same amount of products. The southern negro's farm is worth less than half as much as the southern white man's and he gets from it about two thirds as much. Taking all the farms from our four groups of states and reckoning them according to the value of what they actually produced in 1900 and of their value in 1920, the census ranks them as shown in Table 3.

TABLE 3. THE RELATIVE EFFICIENCY OF WHITE AND NEGRO FARMERS IN THE NORTH AND SOUTH

	Per value of products in 1900	Per value of farms in 1920
Northern whites,	100	100
Southern whites,	51	44
Northern negroes,	49	49
Southern negroes,	34	18

This little table possesses profound significance. It shows unmistakably two types of contrast. First, there is the racial contrast, the result of long inheritance. That, apparently, is what makes the negroes fall below the whites in both the North and the South. There is also a climatic contrast. That, apparently, is why the negroes who come to the North rise above the usual level of their race, while the whites of the South fall below the level of theirs. I realize that the contrast between the two sections is explained in a hundred ways by as many different people. One ascribes it to the fact that slavery was a poor system economically. Another says that the South is cursed for having consented to the sin of slavery. Again, we are told that the predicament of the South is due to the War of Secession, the failure to develop manufactures, the absence of roads and railroads, bad methods of farming, the presence of the negro making the white man despise labor, and many other equally important causes which cannot here be named. Still other authorities ascribe the condition of the South to its supposed settlement by adventurers, whereas the North had its Pilgrims.

I would not minimize the importance of these factors. All are of real significance, and if any had been different, the South would not be quite what it is. All depend upon the two fundamental conditions of race or inheritance, and place or climate. Yet in the contrast between the North and South, the climatic effects seem to be the more potent. Slavery failed to flourish in the North not because of any moral objection to it, for the most godly Puritans held slaves, but because the climate made it unprofitable. In a climate where the white man was tremendously energetic and where a living could be procured only by hard and unremitting work, it did not pay to keep slaves, for the labor of such incompetent people scarcely sufficed to provide even themselves with a living, and left little profit for their masters. In the South slavery was profitable because even the work of an inefficient negro more than sufficed to produce enough to support him. Moreover, the white man was not energetic, and his manual work was not of much more value than that of a negro. Hence, it was easy to fall into the habit of using his superior brain, and letting the black man perform the physical labor. If the Puritans had settled in Georgia, it is probable that they would have become proud slave-holders, despising manual work.

So far as inheritance is concerned, the white southerners, according to the generally accepted principles of biology, must be essentially as well off as the white men of the North. New England has probably had a certain advantage from the strong fiber of her early settlers, but that section is excluded from our comparison because it has so few colored farmers. In New York, Pennsylvania, New Jersey, and the states farther west, the white farmers in 1900 were of highly mixed origin, and there is little reason to think that they inherit any greater capacity than do the white men of the South. Hence, we infer that the difference shown by the census is largely a matter of climate. It has arisen partly by indirect means such as slavery and disease, partly by direct means such as the disinclination to physical exertion. This demands emphasis, for we are told that the South needs nothing but a fair opportunity, plenty of capital, and abundant roads, railroads, and factories, or else it needs only education, a new respect of one race for the other, coöperation between the two for the sake of the common good, and a deeper application of the principles of Christ. All these things are sadly needed, but it is doubtful whether they can work their full effect unless supplemented by a new knowledge of how to neutralize the climatic influences which seem to underlie so many southern problems. In the climate of the South a part of the white population becomes a prey to malaria, the hookworm, and other debilitating ailments. People cease to be careful about food and sanitation. Even those who are in good health do not feel the eager zest for work which is so notable in the parts of the world where the climatic stimulus is at a maximum. Thus one thing joins with another to cause a part of the people to fall far below the level of their race, and to become "Poor Whites," or "Crackers."

These increase in number as one passes from a more to a less favorable climate. It is their run-down, unkempt farms which bring the average of the southern whites so dangerously near the level of the negroes. The best farms of the South vie with those of the North. They show what could be done if all the inhabitants could be instilled with the energy and wisdom of the best.

Aside from North America the only large area where Teutons and negroes come into direct contact as permanent inhabitants is South Africa. There they meet on practically equal terms. The English and Boers began to settle in South Africa in large numbers only in the first half of the nineteenth century. In 1921 the South African Republic contained about 1,500,000 Europeans, 4,700,000 Bantu natives, and 700,000 other natives and Asiatics. A large proportion of the white men were not born there, and hence the new conditions have not had time to produce their full effect. The majority of the natives are Zulus, but the most capable appear to be the Basutos, an allied race who have preserved a large measure of independence in the Drakenberg mountains. Both the Zulus and the Basutos came from the North a few generations ago. Some preceded the white man and some have come since his arrival.

The colored people are most numerous in the north and east of the Republic, that is, in Rhodesia and Natal. The white men are most abundant in the south and in the central plateau, that is, in Cape Colony, Orange River Colony, and Transvaal. With ever increasing force, however, the blacks are pushing into the white man's country. They are brought as laborers for the mines; they are wanted for the farms; they are in demand as servants; and they are themselves taking up farms and successfully cultivating them. They are doing more than this, however, for they are actually ousting the Europeans. In 1902 the English and the Boers finished a bitter war. Ten years later their enmity had almost vanished in the common fear of the negro. Aside from the disturbances due to the European War of 1914, the great political question has long been the black man. One party advocates segregation, with a white man's South Africa in the highlands from Transvaal southward, and a black man's South Africa in Natal and Rhodesia. No black man, they say, should be allowed to live permanently outside his own country, although he might go elsewhere to work temporarily. The other party holds that such measures are too radical, but it also recognizes the gravity of the situation.

The problem presents itself under an economic guise. The colored men have a lower standard of living than the whites. Hence they work more cheaply. They furnish so abundant a supply of labor that white laborers have no chance. Thus a large number of the Europeans—even a tenth according to ardent believers in the future of South Africa—are "poor whites." They are a shiftless set, living from hand to mouth, proud of their race, yet less efficient than the blacks. The problem of preventing them from becoming an immediate charge upon the community is serious. They lack the push and energy which characterize the rest of the white population. According to Stevens, in his book *White and Black*, 5 per cent of the white population in certain regions have fallen so low that they would rather resort to crime than work in competition with the black man. These figures have been questioned, but they are abundantly confirmed by Dr. Andrew Balfour, Director-in-Chief of the Wellcome Bureau of Scientific Research in London. In some lectures on *Sojourners in the Tropics and Problems of Acclimatisation*, published in *The Lancet* in 1923, he states that he "referred the matter to Colonel P. G. Stock, of the Ministry of Health, who knows South Africa intimately, and he confirmed Huntington's statement, pointing out, however, that in parts of the Transvaal chronic malaria may be to blame." The most sinister fact is that these "poor whites" appear to have been largely born in the country. The newcomers are on the whole more energetic. They find employment, and if they have difficulty in one place, move on to another. The poor whites lack the initiative to do this. If they fall into difficulties, they tend to lie down and give up. They need higher wages than the blacks in order to maintain their traditional standard of living. They are not efficient enough to get higher wages. If they had the restless energy which characterizes the children and grandchildren of emigrants from Europe in Canada, for example, they would scarcely fall into such straits.

Since the problem is economic, the South Africans are striving to apply economic remedies. This is wise, but success is doubtful unless other factors are also considered. Back of the economic facts, and in many ways conditioning them, lies the climate. South Africa is supposed to have a climate admirably adapted to Europeans. I shared the common opinion until I began to gather statistics of the effect of climate upon efficiency. These, as will be shown later, indicate that although the South African climate is pleasant, it lacks the stimulating qualities which are so important in Europe and North America. This lack of stimulus increases rapidly as one goes from south to north. Here, then, is the situation that confronts us: In South Africa the white men settled first in the regions most favorable from a climatic point of view and then pushed northward into worse conditions. Even the best parts of South Africa cannot approach England and Holland in the excellence of their climate. Hence, the white settlers are everywhere at a disadvantage. On the other hand, the Bantu negroes have come into South Africa from the north, where the climate is far less favorable than in their new homes. Thus the two races face each other under conditions which lessen the white man's energy, while they stimulate the black man. The whites are still far ahead, and will doubtless continue to be so indefinitely. Nevertheless, the weaker ones are being weeded out and prepared for destruction. What the final result will be, no man can say. It depends upon whether we can discover a means of preventing the deterioration which now seems to attack a portion of the population when people move from a good climate to a worse.

A more striking case than that of South Africa is found in the Bahama Islands. At the time of the American Revolution a considerable number of Loyalists were so faithful to England that they sacrificed their all in order to escape from the new flag with its stars and stripes. Leaving their homes in Georgia and other southern states they sought the British territory of the Bahamas. Other colonists came from Great Britain. Now, after from three to five generations, the new environment has had more opportunity than in South Africa to produce its full effect. Almost nowhere else in all the world have people of the English race lived as genuine colonists for several generations in so tropical a climate. What has been the result? There can be but one answer. It has been disastrous. Compare the Bahamas with Canada. The same sort of people went to both places. Today the descendants of the Loyalists in Canada are one of the strongest elements in causing that country to be conspicuously well governed and law-abiding, and the descendants of other colonists, both British and French, vie with them in this matter. In the Bahamas the descendants of the same type of people show today a larger proportion of poor whites than can probably be found in any other Anglo-Saxon community. Although no figures are available, my own observations lead to the conclusion that the average white farmer is scarcely ahead of the average negro.

Whatever the exact figures may be, there can be no question that in the Bahamas the two races tend to approach the same level. This seems to indicate a marked retrogression of the white race in regions which are climatically unsuitable. Let me hasten to say that many of the more intelligent Bahamans do not differ from the corresponding portions of the Anglo-Saxon race elsewhere. At home they feel themselves handicapped, but when the young people go away to the northern United States or England, they frequently show marked ability. Their inheritance is still good. As to the poor whites, who were described in connection with the lumber industry, it is not so certain that their inheritance remains unimpaired, for in some villages genuine abnormalities both of body and mind are seen. This, however, may be due to the intermarriage of cousins which has been common in certain communities.

The inefficiency of many of the white Bahamans, however, is not due to intermarriage, as is sometimes implied, for villages where this prevails are scarcely worse than those where it is no more common than in America. Nor is the inefficiency due to disease. The hookworm is practically unknown. According to a report of Dr. McHattie, Chief Medical Officer of the Islands, only two cases had been reported up to October, 1913. In this report, for which I am indebted to the courtesy of Dr. J. A. Ferrell of the Rockefeller International Health Commission, the author points out that "the remarkably rapid manner in which the soil . . . dries after even the heaviest rain" prevents the development of the infective larvæ. For similar reasons malaria is no more prevalent than in Delaware, for instance, and in general the islands are decidedly healthful. A monotonous diet may be another detrimental factor, but it is scarcely the root of the matter. Many of the people are well fed, and all could be so if they displayed any energy. Indeed, many people say that life is altogether too easy in the Bahamas. The soil is wonderfully fertile, crops of some kind will ripen at all seasons, and a man can work less than half his time and still readily procure an abundance to eat and wear for himself and his family. On the other hand, we are often told that the difficulties of life have broken the spirit of the inhabitants. The soil, in spite of its richness, is thin, and rocks are so abundant that the plough is almost unknown. Hand agriculture in little patches in the midst of naked limestone is the rule. It cannot be denied that there are difficulties in comparison with many other tropical countries.

For instance, I was talking with a negro whose parents were in a slave ship bound from Africa to Cuba when a British warship captured it. The slaves were taken to the Bahamas and liberated. In answer to a question as to how his parents liked the islands compared with Africa, the son said: "They didn't like it. They used to say, 'In Africa one could lie around all day and do nothing and always find something to eat. Here one has to work or else starve.' " The truth seems to be that compared with North Prussia or Maine the Bahamas are a very easy place in which to make a living, but that much more work is needed than in some other tropical regions. They are at the happy mean. Other difficulties—such as the tropical hurricanes which sweep over the country once in every few years; insect pests, which are neither more nor less harmful than in other countries; the American tariff; competition with Cuba, and above all the isolated position of the islands—are frequently cited as causes of the constant Bahaman failures. The islands are always suffering from bad luck, and something must be to blame.

All these various factors doubtless play a part in retarding the development of the Bahamas. Back of them, however, lies a factor of even greater import, namely, an inertia due to the climate. It does not cause the difficulties mentioned above, but it aggravates them and makes it almost impossible to overcome them. I talked about this with perhaps fifty of the more intelligent people, including both natives and foreigners who have been there a number of years. Almost without exception they said the same thing. "This climate is one of the best in the world. You can see that for yourself. It is very healthful, and we have very few sicknesses. The only trouble is that it does not make one feel like work. In winter it is all right, although even then we cannot fly around the way you Americans do. We always feel lazy, and in summer we want to sit around all the time." As an American picturesquely put it: "Until I came to the Bahamas I never appreciated posts. Now I want to lean against every one that I see." Many of the men and almost all the women complained of feeling tired. Even the children are listless. One young man stated the case very strongly, "We go to bed tired in summer and we get up more tired, and the summer lasts from April to October." Again and again people said: "Oh, it's all very well for you to think we're lazy, but try living here six months or a year and you'll be as lazy as we are. It's something in the air. Just look at these young ministers who come out from England. At first they are full of energy, but after a year or two it oozes out, though their spirit is still as zealous as ever." Two of the ministers spoke of the fact that when they came out they thought nothing of walking twenty miles, but now they dread the thought of two. Several of the most thoughtful and intelligent islanders, men who have succeeded in business and whose judgment would be respected anywhere, said: "We know that we are physically unable to do what English and Americans can do. We are weaker than our fathers, and they were weaker than theirs. It is a grief to send our children away, but in our hearts we know that this is not a white man's country." All this, it must be remembered, is not due to any specific disease, so far as we are aware. Indeed, I met several people who said that a stay of a few years in the Bahamas had improved their health, but at the same time had made them feel inefficient.

Aside from extremely ignorant persons whose opinion is of little value, the only men who spoke of the climate more hopefully were five or six highly trained officials and others occupying positions of authority. These men, without exception, can control their own time. In most cases their office hours are from 9 or 10 a.m. to 2 p.m., or less. They are men of naturally strong physique; they have the opportunity and the will to take regular exercise; and, most important of all, they make long and frequent visits to the United States or England.

The benefit to be derived from a visit to a more bracing climate is astonishing. The contrast between the dull, sallow complexions and thin cheeks of the women and girls who have always lived on the islands and the round, rosy cheeks of those who have recently come back from a long stay at the North is most striking. According to a local saying you cannot tell whether a Bahaman woman is pretty until she goes away, and has a chance to fill out her cheeks and get some color. It is by no means strange that the stronger, more energetic young white people are fast leaving the islands. I asked a Bahaman girl, who had been studying nursing in New York, whether she enjoyed life more in the United States than in the Bahamas. "How can one help enjoying it more there?" she answered. "There one *feels* like doing things. Here one never feels like anything." Like almost everyone else she was sure that it was the climate even more than the new social environment which made the difference.

One thing that surprised me was to hear the Bahamans speak of the stimulus of living in Florida. A native merchant remarked: "If I hire a new man I don't have to ask whether he has been to Florida. I know it by the way he works, but it does not last long." Here again the social environment is an important factor, but various people told me that the air somehow makes them feel more capable of work in Florida than at home. The women of Florida—I heard them say it themselves—are pale and wan compared with their northern sisters. One of them, whose color still shows her northern origin, remarked: "When I come home after a summer in the North, I am full of energy and see all sorts of things that I want to change about the house. But after a month or two I don't care whether things are fixed or not." One hears the same sort of thing everywhere. A factory superintendent from Atlanta, Georgia, told me that the Florida workmen, even the most skillful mechanics, drive him frantic because they are so shiftless and so ready to take a day off whenever they feel like it—far more so than at Atlanta, even though Atlanta seems slow to northerners. Yet, in spite of all these things, Florida is a more stimulating place than the Bahamas.

Its summers are not much better, but its winters are sometimes frosty, while in the Bahamas the thermometer practically never goes below 50° F. Perhaps of greater importance, as we shall see later, is the fact that in Florida the temperature from day to day varies much more rapidly than in the Bahamas, even though both places are in the same latitude. Hence, the mainland is blessed with a genuine climatic stimulus compared with the uniform islands.

The last thing to be said about the Bahamas concerns the effect of the climate on mental activity. Practically all the islanders with whom I talked thought that the effect of the climate on mental activity is at least as great as on physical. Several of the more thoughtful, without any suggestion on my part, put the matter in this way: "The worst thing about this climate is the effect on the mind. Not that people do not have as good minds as elsewhere, but one soon gets weary of hard mental effort. It is extremely difficult to concentrate one's thoughts. At night one cannot seem to make himself read anything serious —nothing but the lightest kind of stories." In our own southern states one hears the same complaint. Even in Virginia the booksellers say that during the long summer almost no one touches a serious book. One feels it everywhere, for on the trains, at the railroad stations, and at the newsdealers it is generally difficult to find the higher grade of magazines. Time and again during a recent journey of three months in the southern states I tried to get such papers as the *Outlook, Independent, Harper's, Atlantic, Review of Reviews, The Century,* and so forth—but all that I could find was trashy story magazines. The dealers rarely keep the better magazines because people will not read them. Lack of training surely has something to do with the matter, but mental inertia due to lack of climatic stimulus seems to be at least equally important.

Let us return now to our question as to a Teutonic and a negro Egypt. The farmers of the northern and southern states, the race problem of South Africa, and the backwardness of the Bahamas, all seem to point to the same conclusion. When the white man migrates to climates less stimulating than those of his original home, he appears to lose in both physical and mental energy. This leads to carelessness in matters of sanitation and food, and thus gives greater scope to the diseases which under any circumstances would find an easy prey in the weakened bodies. The combination of mental inertia and physical weakness makes it difficult to overcome the difficulties arising from isolation, from natural disasters, or from the presence of an inferior race, and this in turn leads to ignorance, prejudice, and idleness. Thus there arises a vicious circle which keeps on incessantly. From its revolving edge a part of the community is thrown off as poor whites, whose number increases in proportion to the enervating effect of the climate and the consequent speed with which the circle revolves. That climate is the original force which sets the wheel in motion seems to be evident, because it is only in adverse climates that we find the "cracker" type of "poor white trash" developing in appreciable numbers. If white men lived a thousand years in Egypt it seems probable that a large proportion of them would degenerate to this type. Whether they would still retain an inheritance of health and mentality sufficient to keep them ahead of a similar body of negroes can scarcely be determined.

The chief reason for doubt in this respect is that we do not yet know just how natural selection would work in such a case. It would almost certainly act in two ways. First, many of the abler young people of the white race would presumably migrate, as they do in the Bahamas. This tendency is generally strongest in the upper classes who can afford to send their children away to school. It is also strong among the young people of all classes who possess more than the average initiative, ambition, and physical strength. It becomes weaker and weaker as one goes down the scale, and almost ceases among the "poor whites," who have so little mental capacity and so much physical inertia that in spite of much grumbling they remain where they are and compete with the colored people. The other kind of natural selection consists of a selective death rate. Children who inherit certain physical and mental traits are more likely to die than are children who do not possess those traits. What the traits are which cause extermination we do not know. A fair skin, a nervous temperament, an excess of activity, and an unwillingness or incapacity to get sufficient rest may be qualities which doom certain white stocks to gradual extinction outside their own climate. In places like South Africa and the Bahamas the temperament which is willing to intermarry with the colored people helps certain types of white people to perpetuate their inheritance, but at the same time it gradually eliminates the qualities of energy, initiative, and inventiveness which seem to be so much more characteristic of Nordics than of negroes.

It must not be forgotten that theoretically it may be possible that some day a carefully controlled series of crosses between whites and blacks may eliminate the weak traits of each and combine the good traits. Thus a race may arise which resembles negroes in its good temper and its capacity to withstand a tropical climate, but which will have the progressive, executive, and inventive capacities of the white race. Such crosses have been made among animals. For example, Mr. M. F. C. Honoré of the Transvaal has sent me the following quotation which he believes to be prophetic of what will some day happen in South Africa. It is from Winston Churchill, the British Cabinet Minister. "At Naivasha [practically on the equator in British East Africa] there is the Government stock farm. One may see in their various flocks, the native sheep, the half-bred English, the three quarter bred, etc. The improvement is amazing. The native sheep is a hairy animal, looking to the unpractised eye more like a goat than a sheep. Crossed with Sussex or Australian blood, his descendant is transformed into a woollen beast of familiar aspect. At the next cross the progeny is almost indistinguishable from the purebred English in appearance, but better adapted to the African sun and climate. It is the same with the cattle. In the first generation the hump of the African ox vanishes. In the second he emerges a respectable English shorthorn."

Such carefully controlled crossbreeding may perhaps be possible among mankind after hundreds or thousands of years. But first we must know what human qualities are "unit characters" so that they are inherited according to the Mendelian law and are not due to the combination of a series of such characters. Then we must learn what qualities are dominant over others so that the presence of one hides the other. Another highly complex problem is to determine what qualities are linked with others so that one cannot be inherited without the other. The fact that linked qualities are very common may mean that certain good qualities like the tolerance of the negro for a hot climate can never be inherited without certain undesirable qualities like the lack of care for the future which is one of the chief causes of negro shiftlessness. Even if such linkage is not an insuperable barrier to the production of a really new race by intelligent crossbreeding there still remains the almost insurmountable obstacle of deep-seated human customs, racial antipathies, and modern ideas of individual liberty. Nevertheless, it is worth while to reflect on the following dream of Lafcadio Hearn: "It is neither unscientific nor unreasonable to suppose the world eventually peopled by a race different from any now existing yet created by the blending of the best types of all races; uniting western energy with far eastern patience, northern vigor with southern sensibility, the highest ethical feelings developed by all great religions with the largest mental faculties evolved by all civilizations; speaking a single tongue composed from the richest and strongest elements of all preëxisting human speech, and forming a society unimaginably superior yet unimaginably un-like to anything which now is or will ever be."

This is an inspiring dream even though most biologists now regard it as impossible. So far as climate is concerned the hard reality seems to be that at present, both by its direct action and through natural selection a warm, monotonous, and unstimulating climate tends to reduce human activity both physical and mental, regardless of race.

CHAPTER III

THE WHITE MAN IN THE TROPICS

THUS far we have dealt with the temperate zone. Even the Bahamas lie north of the Tropic of Cancer. Let us now turn to the torrid zone, which contains the world's richest and most inviting fields of future development. Let us inquire into the effect of that region upon Europeans who attempt to live there permanently. The isolation of the tropical regions, their lack of facilities for transportation, and the great difficulties of agriculture will doubtless be overcome, but that will by no means solve the problem. Two great obstacles will still remain—the native inhabitants and the white man's own mind and body.

Whatever may be the cause, it is generally agreed that the native races within the tropics are dull in thought and slow in action. This is true not only of the African negroes, the South American Indians, and the people of the East Indies, but of the inhabitants of southern India and the Malay peninsula. Perhaps they will change, but the fact that the Indians both of Asia and South America have been influenced so little by from one to four hundred years of contact with the white man affords little ground for hope. Judging from the past, there is scant reason to think that their character is likely to change for many generations. Until that time comes they will be one of the white man's greatest obstacles. Experience shows that the presence of an inferior race in large numbers tends constantly to lower the standards of the dominant race. Here in America, although the negro forms only a tenth part of the population, he is one of our gravest problems. Yet he is not so great a handicap as are the native races of the tropics. Whatever the negro may have been when he was first brought to America, he is now less stolid and indifferent, more subject to stimulating influences than he was when he came, or than the Indians of tropical America. It is literally true in South America, for instance, that the more an Indian is paid the less he will work. If one day's pay will buy two days' food, he will work half the time; if the pay is increased so that one day's pay will buy food for three days, he will work one third of the time. The experiment has been tried again and again. The most considerate employers of tropical labor agree with the most inconsiderate that in general it is useless to attempt to spur the Indians by any motive beyond the actual demands of food and shelter. Kindness and consideration on the part of the employer undoubtedly promote faithfulness, but they seem rarely to arouse ambition or energy. With the negro in Africa, as everyone knows, much the same condition prevails, but where he has been brought to the United States this is by no means so true. For example, in Central America it is generally thought that a negro from Jamaica is more efficient than an Indian, while a negro from the United States is much more efficient. The negro in the United States is generally considered more efficient than he was in Africa, whereas his stay-at-home brother and the Indian of tropical America, remaining in their old environment, do not seem to have changed.

Doubtless the change in the negro is due to a new social environment quite as much as to a new physical environment, and many authorities believe that the social change is the more important. This, however, does not materially alter the case. As conditions are now, it is extremely difficult to change the physical environment of tropical races so long as they remain in their present habitat, and it seems to be equally difficult to change their social environment. Those who dwell permanently in the white man's cities are influenced somewhat, but the results are often disastrous. Here as almost everywhere within twenty degrees of the equator, the general tendency seems to be to revert to the original condition as soon as immediate contact with the white man is removed.

This does not mean that contact with a higher civilization will never benefit the people of the tropics, but merely that the process is bound to be slow. The aborigines of tropical America, for example, show little sign either of disappearing, or being swallowed up by a multitude of immigrants, as has been the case in temperate latitudes, nor do they appear to be changing their character. On the contrary, in Latin America, the only tropical region except Australia where the white man has settled in large numbers, the proportion of Indian blood is apparently increasing at the expense of the white, and the Indians act and think almost like their ancestors three or four centuries ago. This is largely because the white man, except in a few favored places, suffers from tropical diseases more than does the native, and his children tend to move away if strong, or to be weaklings who die young and leave few children. It is notorious that India contains almost no fourth generation of Indian-born British. The British children are either sent back to Europe to recover their health, or else become enfeebled and their descendants die out. Even with the help of modern medical science, it is not yet certain that the permanent white population can increase greatly, although sojourners are sure to become numerous. In Australia, to be sure, the white man seems to be succeeding within the tropics, but he is still new there and has the inestimable advantage of active natural selection and freedom from contact with the natives.

In many well-populated tropical countries modern science lowers the death rate among the natives, and thus increases their numbers. The white man has permitted the native population of India to double and that of Java to increase seven-fold, partly by conquering diseases and partly by the prevention of famine and war. If the conclusion just reached is correct, it seems probable that tropical countries will long continue to maintain a dull, unprogressive population. Contact with such a population constantly exposes the white man to a most deteriorating influence. For example, the inferior mental ability of the lower race, and its incapacity for effective organization lead to the abuse of its labor and to its exploitation in some form of peonage, even though the fact may be disguised by legal phraseology. Again, the presence of a despised race is almost certain to lead to low sexual morality. In the same way, political equality becomes a mere form of speech, for the dominant race will not permit the other to gain rights at its expense. Manual labor, too, is despised, for it is associated with the idea of an inferior race. All these things may be looked upon as disadvantages of the lower race, but I believe that the higher reaps by far the greater injury. The conditions just mentioned appear to be among the most potent factors in rendering it difficult for the white man to attain as much success in tropical regions as in those farther to the north or south. Their evil effect is roughly proportional to the difference between the two races. That difference is at a maximum where a low tropical race remains in its original, unstimulating environment, and is brought into contact with immigrants of a highly developed race who completely change their environment. The newcomers are released from old restraints at a time when they stand in peculiar need of them. Instead of being stimulated to greater love of political freedom and equality, sterner morality, and more intense industry, as was the case among the settlers in New England, the immigrants are in danger of being weakened in all of these respects. The effect on the original immigrants is bad enough, but on their children it is far worse. The settler, or European colonist, who is possessed of wealth and power, can to a slight degree shield his family, but even in such cases the children are in constant contact with servants. They grow up with a supreme contempt for the natives, and at the same time with the feeling that they can treat them as they choose. If poorer people, that is, colonists in the ordinary sense of the word, attempt to live in the tropics, especially if they are people who work with their hands, their children are exposed still more to all the contaminating influences of contact with the natives. Hence, the second and third generations, and the fourth and fifth, if there are any, suffer more than their ancestors.

The degree to which the indirect or external handicaps of tropical countries are effective in lowering the standards of civilization depends largely upon the amount of energy and will power possessed by the inhabitants. This, in turn, depends upon physiological conditions. Obviously, diseases have much to do with the matter. This subject has been so much discussed that I shall here refer to it only briefly. There can be little doubt that malaria and the many other diseases which are characteristic of tropical countries play an important part in causing a low state of civilization. The old idea that the people who live in tropical regions are immune to local diseases is no longer accepted by students of tropical medicine. Adults, to be sure, are often immune, but apparently this is only partially true of children. Vast numbers of children die in infancy and early childhood from the diseases which prevent the white man from permanently living within the tropics. Others suffer, but recover. They bear the results with them to the grave, however, in the form of enlarged spleens, or other injuries to the internal organs of the body. The world has of late been astonished at the ravages of pellagra and of other diseases due to such organisms as the hookworm. People who are subject to them cannot be highly competent. Their mental processes, as well as their physical activity, are dulled. So long as a community is constantly afflicted with such disorders, it can scarcely rise high in the scale of civilization. Nothing is more hopeful for the tropics than the rapid progress in the control of these diseases. If they could be eliminated, not only might the white man be able to

live permanently where now he can be only a sojourner, but the
native races would probably be greatly benefited. How great
this benefit would be we cannot yet tell, but the elimination of
the diseases which especially affect children would probably do
much to increase vitality, energy, and initiative. This in itself
would be an immeasurable boon not only to the natives, but to
the white man, who would thereby be freed in part from some of
his worst social dangers.

This highly desirable result cannot be obtained quickly. The
achievements of the United States in Panama are sometimes said
to prove that diseases can be eliminated anywhere in tropical
countries. This is true, but it must be remembered that Panama
is a highly specialized case. During the building of the Canal
a great number of people were collected in a small area, and
enormous sums of money were freely expended. Everyone was
subject to strict, semi-military rule, and similar conditions still
continue. Such methods cannot be applied to millions of square
miles. The expense would be prohibitive. The ordinary farmer
in tropical regions cannot expect to be protected by his government. He must protect himself. In the long run even tropical
races may learn this lesson, but it will be a difficult and expensive
task, and will require a radical change in the people themselves.
Such a change will doubtless come, but not for generations, and
not until a long selective process has gone on whereby those
who do not adopt modern medical methods will gradually be
eliminated, while those who adopt them will persist.

There has been so much misunderstanding of Panama and so
many wild statements that it may be well to set forth the exact
facts. The Health Department of the Panama Canal, as it is now
called, has charge of three districts, whose population in 1917
was as follows: the city of Panama, 61,074; the city of Colon,
25,386; and the Canal Zone, 27,543. For purposes of health and
sanitation all are under the control of the United States, and no
expense is spared to make them as healthful as possible. In order

to avoid the complications due to the influenza epidemic of 1918, let us take the period from 1912 to 1917. By 1912 the health measures of the United States Army had reached such perfection that the death rate had been reduced 50 per cent. The improvement still continues, but it is now slow and apparently does little more than keep pace with the similar improvement in the advanced parts of the world. The two cities of Panama and Colon contain the ordinary mixed population of tropical seaports: negroes from the West Indies; Mestizos—half Spanish, half Indian—from the neighboring parts of Central and South America; a few Chinese and other Asiatics; some Europeans and Americans. A considerable number of the employees of the Canal live there. The Canal Zone, on the other hand, contains a large proportion of Canal employees, chiefly Americans, West Indian negroes, and Europeans. Among all of these the percentage of men between twenty and fifty years of age is large. The following figures show the crude death rates from 1912 to 1917 among the civilian population, excluding soldiers, in the three districts of Panama and in certain other areas with which comparisons may profitably be made.

Panama	. 30.5	Chile	. . . 27.9	Bombay (1910-1912)	. . 37.0
Colon	. . 24.8	Spain	. . . 22.0	Calcutta (1910-1912)	. . 26.1
Canal Zone	. 13.6	United States	. 13.9	Amsterdam (1901-1913)	. 12.6

A multitude of other figures might be presented all of which would show that while the work done in Panama has been admirable, the general conditions of health in the cities of Panama and Colon are still twice as bad as in the advanced parts of the world. They are about on a par with those of similar cities of India, for Bombay and Calcutta, by reason of their size and desperate overcrowding, presumably have higher death rates than do Indian cities as small as Panama and Colon. The death rate for infants under one year bears out this general conclusion, as appears from the following figures showing the deaths per one thousand births in 1915, 1916, and 1917:

The area where births are registered in the United States includes only a small part of the South, so that the death rate among colored infants as a whole is higher than appears above. In Richmond, Virginia, during 1917, 1918, and 1919, it averaged 198. In cities farther south it doubtless reaches a level as high as that for people of all sorts at Panama.

The foregoing data make it obvious that the widespread idea as to the healthfulness of Panama is based solely on the small number of people in the Canal Zone. But the death rate of 13.6 given above for the Canal Zone has by no means the significance that is usually supposed. Its use for comparative purposes is vitiated by two facts; first, the number of deaths by violence, chiefly by accident, is unusually high in the Canal Zone; and second, the inhabitants of the Canal Zone are a highly selected group mostly of good physique and in the prime of life and hence bound to have a relatively low death rate no matter where they live. The best way to make a fair comparison is to take people of the same age, sex, and occupation who have otherwise also been selected by the same method, and compare the death rates in different places. But this is impossible. As the next best thing let us take the death rate from 1912 to 1917 among the Canal's white employees from the United States and compare it with the rate for men of similar age elsewhere. If we assume that the proportion of white men of different ages in the Canal Zone is the same as among the white employees from the United States and was also the same at the census of 1920 as in the period from 1912 to 1917, both of which are essentially the case, it is easy to compute the relative death rate in other regions on the same basis as the rate for Panama. Using the data prepared by the International Institute of Statistics together with the records of Yale University, we find that if the proportion of men of various ages were the same in the other places as among the white American employees of the Panama Canal, the death rates *among such men* would be as follows:

	Death rate from all causes	Approximate death rate when deaths due to violence are eliminated
New York State, 1906-1915	12.3	10.6
Connecticut, 1906-1915	10.8	9.3
Washington State, 1908-1913	7.8	5.3
New Zealand, 1906-1915	6.7	4.8
White Canal employees from the United States, 1912-1917	4.9	2.7
Students in Yale University, 1912-1917 . .	1.9	1.7

Does this mean that the climate of the states of New York and Connecticut is relatively bad, while that of Panama and the home of Yale University at New Haven is remarkably good? Not at all. It simply means that the two states have relatively normal death rates for their particular climates and for a comparatively unselected population. They are handicapped by their numerous unhealthful factories and cities and by the great number of their immigrants, many of whom are poor, ignorant, and of low caliber mentally. Moreover, many of the more energetic young people have migrated westward. The adventurous and persistent qualities which lead to migration are partly due to health and physical vigor, and partly to mental initiative, adaptability, and readiness to try a new life and new methods. It needs no demonstration to show that such people are sure to have a low death rate, especially when they are highly prosperous, as in the state of Washington. They have the physical vigor to withstand disease, they have the good sense to take care of themselves, and they have the means wherewith to purchase good food, good shelter, good sanitation, and good medical service. Since New Zealand is harder to reach than Washington, its immigrants

have been even more highly selected for thrift, health, and
physical and mental vigor.

Panama, like Washington and New Zealand, attracts chiefly
the more vigorous type of people. The man who is organically
diseased rarely thinks of going there. Moreover, in Panama
white employees come from America as adults. On the contrary,
many of the people of Washington and New Zealand were born
there and have remained regardless of whether they possess the
pioneer vigor and initiative of their parents. Again, even if they
have the brave spirit that overcomes physical handicaps, the organically weak are not allowed to go to Panama as employees.
If they try to go, they are weeded out by physical examinations.
Even that, however, does not end the matter, for the examinations are repeated each year. Every individual who shows signs
of weakness is advised to leave Panama as soon as possible;
many are ordered home; and not a few are deported, especially
those suffering from mental disorders. During the three years
for which I have been able to find data (1914, 1915, and 1917)
the deportations on account of disease among employees of all
sorts, both white and colored, amounted to approximately 40
per cent of the total deaths from disease. If these people had
stayed in Panama, as they stay in New York, Connecticut,
Washington, or New Zealand, many of them would soon have
died. Inasmuch as such deportations have been going on for
years, it is practically certain that without them the death rate
at Panama would be decidedly larger than now. In addition to
the persons who are deported, a far larger number go home
voluntarily on the advice of their physicians. Moreover, many
who show no immediate signs of disease remain at home after
one of their earlier furloughs because they find the climate at
Panama uncomfortable.

In addition to all this it must be remembered that the white
employees at Panama are practically all officials or clerks; they
belong to a class of society which by reason of its intelligence is

able to take care of itself, so that its death rate is normally much lower than that of the great body of men of similar age. Moreover, the employees are well paid and admirably housed. They likewise have long and frequent vacations at home whereby the effect of the tropical climate is partially neutralized. All these conditions, even without the excellent medical care which the employees receive, free of cost, would insure a degree of health in Panama much better than in most tropical regions. On the other hand, if the population of Panama were an ordinary unselected type and if none of the weak or sick were sent away, it seems probable that in spite of admirable sanitation and medical care the death rate would be larger than in New York among people of similar age. This last statement is merely an opinion, since by its very nature it is not susceptible of actual proof. We know as a fact, however, that the death rate at Panama is greatly lowered by the selection of healthy, intelligent employees as well as by good medical care.

The conditions at Yale or any other university suggest that in such cases selection is even more important than medical care. From June, 1912, to June, 1917, the average number of undergraduates at Yale was 2476, among whom the total number of deaths was nine through disease and one by accident. This gives an annual death rate of 0.8 per thousand against 4.5 for all the young men of Connecticut between 15 and 24 years of age. In other words, an unselected young man in Connecticut is 5.6 times as likely to die as is a selected Yale student. If a similar ratio prevailed among the University men up to the age of about 55 years, and if the proportion of men at each age were the same as at Panama, their death rate, barring accidents, would be only 1.7 against 2.7 among the white American employees at Panama. Now as a matter of fact, from the standpoint of health the employees at Panama are far more rigidly selected than are the Yale students. No medical examination is required for entrance to the University, no one is actually sent away because of his health, and the amount of medical attention is less on the whole than at Panama. Other things being equal, this ought to give a higher death rate at Yale than at Panama. Yet, as a matter of fact, it actually gives Panama the higher rate by 60 per cent. This turning of the tables against Panama seems to be due to the adverse climate.

The net result of the preceding investigation is this: There can be no doubt of the great value and success of the medical and sanitary work at Panama. It has cut the death rate in halves at the cities of Panama and Colon. Nevertheless, the death rate in those cities is still extremely high, about twice that of the United States as a whole. So far as the white people at Panama are concerned, the death rate is very low, but that proves nothing about the climate. It merely proves that it is possible to obtain practically any death rate by selecting the cases. One could go to hospitals and select the critical cases. That might give a death rate of eight or nine hundred. One might select college athletes and from time to time throw out any who showed signs of illness, and the death rate would be zero. But to use such death rates as evidence concerning the climate would be highly misleading. It is poor policy to use any such reasoning in respect to Panama, Northern Australia, or any other region where the climate possesses disadvantages. To do so encourages false hopes. When these are disappointed, people tend to blame the whole science of tropical medicine. That science is doing wonderful things, but as yet there is no evidence that it has overcome the effects of climate, although it has certainly mitigated them. We shall return to this subject in connection with Australia. There, as in Panama, the tropical death rate is lower than those of better climates, but this is due primarily to the selection of certain types of residents. I have dwelt on this matter because there is a vast deal of misapprehension and very little realization of the importance of selection.

Let us return now to our main question. Suppose that the

white man should succeed in cultivating the tropical forests, transversing the waste places, and conquering the diseases. Suppose also that he should eliminate the deteriorating influences of low social and moral standards among the natives. But suppose also that there were no selection of the white colonists. If all this were suddenly done, and average unselected white men were set down in a tropical garden of Eden, would they be able to hold their own among the peoples of the world? Would Teutons or Latins under such circumstances be able permanently to maintain as high a standard of civilization as is maintained by their brothers in Europe? Or would there be a change in some of the traits which we are wont to call racial? Clearly we are back at the point where we started, and are confronted by the question of race versus place. We must determine how much of our European and American energy, initiative, persistence, and other qualities upon which we so much pride ourselves is due to racial inheritance, and how much to residence under highly stimulating conditions of climate.

One of the lines along which we may seek for an answer is by a comparison of the character of Europeans in tropical countries with their character in the temperate zone. Whatever differences we may find are presumably due partly to physiological and partly to sociological causes, but they manifest themselves chiefly through the will. In tropical countries weakness of will is unfortunately displayed not only by the natives, but by a large proportion of the northerner sojourners. It manifests itself in many ways. Four of these, namely, lack of industry, an irascible temper, drunkenness, and sexual indulgence are particularly prominent, and may be taken as typical. Others, such as proneness to gambling and disregard for the truth, might equally well be considered if space allowed.

In the quality of industry the difference between people in tropical and other countries is well known. We have already touched on it in the Bahamas, but let us amplify it further.

Practically every northerner who goes to the torrid zone says at first that he works as well as at home, and that he finds the climate delightful. He may even be stimulated to unusual exertion. Little by little, however, even though he retains perfect health, he slows down. He does not work so hard as before, nor does the spirit of ambition prick him so keenly. On the low, damp seacoast, and still more in the lowland forests, the process of deterioration is relatively rapid, although its duration may vary enormously in different individuals. In the dry interior the process is slower, and on the high plateaus it may take many years. Both in books and in conversation with inhabitants of tropical regions one finds practical unanimity as to this tropical inertia, and it applies both to body and mind. After long sojourn in the tropics it is hard to spur one's self to the physical effort of a mountain climb, and equally hard to think out the steps in a long chain of reasoning. The mind, like the body, wants rest. Both can be spurred to activity, but this exhausts vitality. The common explanations of tropical inertia are diverse. One man says that within the tropics hard work is unnecessary, because salaries are high; another asserts that it is because servants are cheap; still another claims that hard work is dangerous to the health; and almost all agree that "anyhow, one doesn't feel like working down here." Probably all four of these factors coöperate, and each, doubtless, produces pronounced results, but the last two, health and "feeling," seem the most important.

In spite of individual exceptions, white men who spur themselves to exert their minds as earnestly and steadily within the tropics as at home are in great danger of breaking down in health. They become nervous and enfeebled, and readily succumb to tropical diseases. This is one of the most powerful deterrents to the development of an efficient white population in equatorial regions. If the more intellectual members of the community ruin their health, they are almost sure to die before their time, or else to go back to the North. In either case they are not likely to leave many children to perpetuate their characteristics. Thus if white colonization takes place on a large scale within the tropics, there is grave danger that the physically strong but mentally lethargic elements will be the ones to become the ancestors of the future population. In the past this factor must have operated to weed out the more intellectual members of each of the many races that have migrated toward the equator. The inertia which prevents the less competent members of a tropical community from overworking may perhaps be interpreted by teleologists as a merciful provision of Providence to warn man that he must not work too hard in the torrid zone, but that will scarcely help to advance civilization. Few people will question the reality of the tropical inertia. It is the same lassitude which everyone feels on a hot summer day—the inclination to sit down and dream, the tendency to hesitate before beginning a piece of work, and to refrain from plunging into the midst of it in the energetic way which seems natural under more stimulating conditions.

Lack of will power is shown by northerners in tropical regions not only in loss of energy and ambition, but in fits of anger. The English official who returns from India is commonly described as "choleric." Every traveler in tropical countries knows that he sometimes bursts into anger in a way that makes him utterly ashamed, and which he would scarcely believe possible at home. Almost any American or European who has traveled or resided within the tropics will confess that he has occasionally flown into a passion, and perhaps used physical violence, under circumstances which at home would merely have made him vexed. This is due apparently to four chief causes. One is the ordinary tropical diseases, for when a man has a touch of fever, his temper is apt to get the better of him. In the second place, the slowness of tropical people is terribly exasperating. The impatient northerner uses every possible means to make the natives hurry, or to compel them to keep their word. His energy is usually wasted—

the native remains unmoved, and the only visible result is an angry and ridiculous foreigner. Yet a show of anger and violence often seems to be the only way of getting things done, and this is frequently used as an excuse for lack of self-control. In the third place, the consequences of becoming angry are less dangerous than elsewhere. The inert people of tropical countries often submit to indignities which an ordinary white man would bitterly resent. Of course they object to ill treatment, and will retaliate if possible, but they generally do not have sufficient energy or cunning to make their vengeance effective against the powerful white man. Finally, those who have lived in the tropics generally find that, even when things go smoothly, and they are in contact with people of their own kind and are in comparatively good health, they are more irritable than at home. In other words, their power of self-control is enfeebled. Of course there are many exceptions, but that does not affect the general principle.

Drunkenness, our third evidence of lack of self-control, need scarcely be discussed. Within the tropics, the white man's alcohol in the form of rum is scarcely more injurious to the natives of Africa than it is in other forms to himself. In places such as Guatemala and parts of Mexico, drunken men and women may be seen upon the streets at almost any time of day. Nowhere else, during extensive travels in America, Europe, and Asia, have I seen so much drunkenness as in Guatemala. Among white men a large number drink as badly as the natives. Here is an example; a railway conductor was telling me about drinks in Guatemala:

"They've got something here called 'white-eye,' " he remarked. "You know that Mexican 'mescal,' and how strong it is? Well, white-eye's got mescal chained to a telegraph pole. Yes, I drink it. A man's got to drink something. The first time I tried it, I got crazy drunk and smashed things up the way they all do. I was arrested and fined fifty dollars. [This is really only two and a half, for Guatemalan currency consists of non-redeemable

paper, which at that time was worth about five cents on a dollar
—a characteristic evidence of tropical incapacity.] I got fined
several times that way and didn't like it. Then one day when I
was going to get drunk, I said to myself, 'I'll go and pay my fine
now and then they won't bother me.' I did that several times, and
the 'jefe politico' liked it [presumably because it was an easy
way of pocketing the money]. Then he said to the police: 'Don't
bother this man. Just let him get drunk all he likes, and he'll pay
his fines at the proper time.' I tell you, white-eye is bad stuff.
The only proper way to drink it is to take a quart bottle in the
morning. Find a place that will stay shady all day. Drink the
whole thing right down and get so dead drunk that you will
sleep till night."

I do not cite this man as typical of all the white men in the
tropics. Far from it. Many conduct themselves with sobriety
and industry, but such men almost invariably make frequent and
protracted visits to the better climate of the North. If a white
man stays steadily for long periods in the tropics, however, and
if his character has any weak spots, they are almost sure to be
exaggerated. The drunkenness of the tropical white man arises
in part from the constant heat, which makes people want something to drink at all times, partly from the monotony of life,
and still more from the absence of the social restraints which
exercise so powerful an inhibitory influence at home. Back of all
these things, however, among both the white men and natives,
there seem to lie two conditions which are directly connected
with the climate. One is the same enfeeblement of the will which
makes a man burst into anger. The other is a constant feeling
of inefficiency which makes a man crave something to brace
him up.

The last of the ways in which weakness of will is evident in
tropical countries is the relation of the sexes. Its importance can
scarcely be overestimated. It leads to the ruin of thousands of
northerners, even though they do not yield to drink, to anger,

or to laziness. When once they have fallen into pronounced immorality the other weaknesses soon follow. The condition of the native races is still worse. Everywhere within the tropics missionaries say that their converts can be taught honesty, industry, and many other virtues, but that even the strongest find it almost impossible to resist the temptations of sex. Many Europeans condone this. They say that it is natural, and that the natives had better be left to their own conventional ways of restricting but not preventing sexual intercourse. Perhaps they are right, although I cannot be certain. That is not the point, however. We are at present concerned with the effect which free indulgence has upon civilization and upon the capacity for progress. This may be illustrated by what Gouldsbury and Sheane, for example, say of the Zulus in northern Rhodesia. They hold that one of the greatest reasons why these people remain so backward is that their thought and energy are largely swallowed up in matters of sex. During the years when the young men ought to be getting new ideas and thinking out the many little projects and the few great ones which combine to cause progress, the vast majority are thinking of women, and planning to gain possession of some new woman or girl. Under such circumstances no race can rise to any high position.

The causes of these conditions are various. Many writers dismiss the matter by saying that the social standards of tropical people are low and tend to cause northerners to conform to them. This is true, but it explains nothing. A real, though minor, reason for the lowness of the standards is found in the free, open life which is almost universal within the tropics. People are out of doors so much and it is so easy to meet in secret that temptation arises very frequently. Much more important is the scanty dress of the women, and its character, which calls attention to their sex. Livingstone speaks with disgust of the way in which his carriers, hour after hour, discussed the breasts of the half-naked women whom they met. Even in the North women seem to be strangely indifferent to the effect of their mode of dress upon men. They do not seem to think that they are responsible if their low-necked gowns and the making of their clothing in such a way that each little movement of their bodies can be detected, stir men's passions. They appear oblivious to the fact that the display of their beauty often means that some other woman must pay the penalty. Within the tropics these conditions are exaggerated. I believe I am speaking within bounds when I say that any young man of European race with red blood in his veins is in more danger of deteriorating in character and efficiency because of the women of the tropics than from any other single cause.

The strength of this deteriorating force is not merely external. Either the actual temptation to sexual excess is greater in the tropics than elsewhere, or else the inhibitory forces are weakened by the same processes which cause people to drink to excess, to become unduly angry, and to work slowly. Hellpach states that it is said that in southern Italy sexual irregularities increase greatly at times when the hot, damp wind known as the sirocco blows across the Mediterranean from the deserts of northern Africa. This is so well recognized among the people themselves that offenses committed under such circumstances are in a measure condoned. Violence, too, is more common at such times, for self-control of every kind is weakened. In eastern Turkey the hot desert winds cause the whole community to become cross and irritable. I have there seen a missionary, a man of unusual strength of character, shut himself up in his study all day, because he knew that he was in danger of saying something disagreeable. I cite this case because, among the people whom I have known, missionaries are, on the whole, most completely masters of themselves and the least likely to let minor circumstances turn them from the Christlike lives which they are striving to live day by day before the native communities. For this same reason, to return to our immediate subject, I quote the

remark of a missionary in Central America when we were discussing the morality of the country. He was a most austere man,
a member of a small and extremely devout sect, and his whole
being was devoted to preaching the gospel. Speaking of his own
experiences, he said:

"When I am in this country, evil spirits seem to attack me.
I suppose you would call them something else, but that is what
I think they are. When I am at home in the United States I feel
pure and true, but when I come here, it seems as if lust were
written in the very faces of the people."

In all the evils which have just been mentioned—laziness,
anger, drunkenness, and immorality—social causes undoubtedly
play an important part. A strong public opinion would save
many a young northerner from drink and immorality, and would
keep him faithful to his work. A clear religious faith or a high
ideal of duty would do the same thing. Good homes, proper dress,
and many other material changes would help greatly. So, too,
would a study of how it has come to pass that certain tropical
races, in spite of their environment, have developed comparatively high moral codes to which they strictly adhere, while a
few have actually learned the lesson of industry. Along with the
social aspect of the question, however, and neither more nor
less important, goes the physical. We must discover to exactly
what extent physical conditions help or hinder the development
of strong character. That is the purpose of the chapters that
follow.

CHAPTER IV

THE EFFECT OF THE SEASONS

IN comparing Teutons with negroes, or tropical people with those of the temperate zone, we have been following a method as old as the days of Aristotle. Such comparisons have led to most interesting generalizations not only at the hands of Aristotle himself, but of many other men such as Montesquieu, Humboldt, and Rätzel. Yet the importance of climate as a factor in civilization is still in doubt. For instance, no one denies that South Africa is confronted by a grave race problem, but many say that it is purely economic, and has nothing to do with climate. They support this view by strong arguments. Thus we are left in uncertainty. The only way to remove this is to devise some method whereby to separate the effects of climate from those due to all other causes, whether economic, historic, social, religious, racial, or something else. Accordingly, the rest of this volume will be devoted to an investigation of the exact effect of various climatic factors upon selected groups of people, and to an attempt to discover how human energy and other qualities would be distributed if all the earth's inhabitants were influenced like these particular groups.

In the study of climate one of the most puzzling features is the diversity of opinion among persons of good judgment. For instance, at what season do people work fastest in the northern United States? Some will say the winter, some the spring, a considerable number the fall, and a few the summer. Most will say that they are least efficient in summer, but others believe that they are at their worst in the early spring or late winter.

Again, ask a dozen friends whether they work best on clear days or cloudy. The majority will probably answer that the first clear day after a storm is by all means the best. A small number will perhaps think the matter over more carefully, and then say that after a storm the clearness of the air and the brightness of the sun are certainly inspiring, but one really accomplishes more when it rains.

This divergence of opinion is due largely to the fact that climatic effects are of two kinds, psychological and physiological. We are always conscious of the first, but often unconscious of the second. The two are admirably distinguished in Hellpach's book on *Geopsychische Erscheinungen.* An example will make the matter clear. It is well known that at high altitudes the number of red corpuscles in the blood increases enormously, and the capacity to absorb oxygen and to give out carbon dioxide is correspondingly modified. Yet many people can go to altitudes of 5000 feet or more without realizing that their physiological functions have been altered. To cite my own case, up to the age of twenty-one I had never been a thousand feet above the sea. Then I went to live at an altitude of 4500 feet. The only physiological effect of which I was conscious was unusual sleepiness for the first few months, but whether this was due to the altitude or to the dryness of the air, I do not know. For two or three years I never thought of the physiological effect of the altitude until one day, happening to have climbed to a height of 7000 feet, I began to run up hill. I lost my breath and became tired so quickly that I was alarmed and thought I must be sick. I was much relieved when it occurred to me that the altitude was not favorable for running up hill. Manifestly my physiological functions were different from what they were at sea level, although I was unconscious of it. On the other hand, psychologically I was daily conscious of living in a place where the air was extraordinarily clear, and where the mountains were always in sight across a splendid plain twelve hundred feet below us. Presumably both the physiological and psychological conditions had an appreciable effect upon the work of every day, but which was the greater it is impossible to tell.

In this connection Lehmann and Pedersen state an interesting fact. In Denmark and Norway they made a series of daily tests of the strength of three individuals by means of a dynamometer. They found that the change of atmospheric pressure due to an ascent of two or three thousand feet makes no appreciable difference. A similar descent, however, is accompanied by a marked increase of strength which disappears within three or four days. They suggest that this may be due to the persistence of abundant red corpuscles when people come down from high places. The red corpuscles multiply very rapidly under the influence of low pressure, but are slower in disappearing when the pressure once more increases. Thus, for the first day or two after a person has come down from the mountains, more than the normal amount of oxygen may be absorbed, and muscular strength correspondingly increased. Possibly this is why mountaineers are generally so irresistible when they descend upon the plains in sudden raids. My colleague, Prof. H. E. Gregory, suggests that this may account for the fact that in the horse-races of the pioneer days of the southwestern United States, the poor, scrawny animals brought down from the mountains by the Indians usually belied their appearance and outran the better-looking animals of the white men. They may have had an excess of red corpuscles. Professor Gregory adds that in some of the highland regions of South America there is a strict rule that before a race the competing horses must spend a certain number of days at the race course. This may have arisen because the animals which race directly after coming from the mountains are apt to win. There is considerable doubt as to the truth of this theory, but it illustrates the possibility that we may be deeply influenced by atmospheric conditions of which we are almost unconscious.

In our opinions as to the effect of the seasons or of daily changes of weather the relation between psychological and physiological influences is probably the same as in the case of altitude. The external conditions which we see and feel make a greater impression than those which prevail within our bodies. For example, most of us think that in the northern United States we work fast in winter. As a matter of fact, the statistics of ten thousand people show that we work slowly. The ordinary impression is apparently psychological. In order to keep warm out of doors in winter we walk fast and this leads us to think that we do everything rapidly.

Again, the blue sky, clear air, bright sunshine, and fresh colors of the first day after a storm are unquestionably inspiring, but does that inspiration make us work any better? May it not lead to a nervous excitement which actually hinders our work, by causing us to look out at the beauties of nature or to be less concentrated in other ways? The actual figures show that, taking the year as a whole, on dull days, especially the second such day when a storm begins to clear, we accomplish more than on bright days, even though we grumble about the clouds and the dampness. A bright day certainly makes us cheerful, but its chief helpfulness, so far as our work is concerned, is felt when it is a change from the monotony of a series of dull days. Clouds and rain produce exactly as much rejoicing when they succeed prolonged clear weather of the kind that we praise so highly. In America I have never seen so much rejoicing over a bright day as I have seen in Turkey when the first rain fell after the long subtropical summer with its truly superb weather. The rejoicing was in part due to the fact that the coming of the rains means good crops, but I have again and again seen exuberant joy among people to whom the crops made no difference whatever. I have seen Americans shout for joy because the clouds had come, and run out into the rain to let the cool drops refresh their faces.

The questions which have just been asked and the possibilities that have been suggested show how indefinite are our ideas of the effect of climate. We understand its psychological effects fairly well. We know little of its physiological effects, however, except when they are extreme or unusual, or when people are sick or are in some other pathological condition. We need to determine how ordinary people are influenced by ordinary conditions of weather. That is the purpose of our present discussion. The most feasible way to do this, as has already been said, is to take groups of people who live in a variable climate, and measure their efficiency under different conditions of weather. The best and fullest test of efficiency is a person's daily work. If the subject does not know that he is being tested, so much the better. Piece-workers in factories are doing exactly what is required for our purpose. Accordingly, to begin with New England, I have taken the daily records of about 300 men and 250 girls,—most of them for a complete year. The records are distributed over the four years from 1910 to 1913. The 550 people were employed in three factories in the cities of Bridgeport, New Britain, and New Haven, in the southwestern part of Connecticut. In all cases the officials in charge of the factories were most courteous and helpful in assisting me to obtain the necessary data, and I wish most warmly to express my gratitude to all concerned.

In the selection of operatives for such a purpose, various conditions must be fulfilled. In the first place, they must be piece-workers who are paid according to their work and not at a fixed rate per day. In the second place, they must be employed in factories where their output is not limited by restrictions imposed by unions, or by the fear that if they earn too much, wages will be reduced. They must be doing work that is of essentially the same kind every day, so that their wages will not vary much because they are sometimes engaged upon new and unfamiliar tasks, or upon easy tasks at some times and hard ones at others.

Furthermore, the same people must work steadily for month after month, throughout the year, if possible, and without taking much time off, as is such a common practice among factory hands. Finally, they must be working where there is abundant incentive to steady, faithful work, where the conditions of air and light are reasonably good, and where accurate daily records make it possible to determine not only the daily wages of each individual but the average efficiency per hour or per day of standard length. The number of factories where all these conditions are fulfilled is small, for they demand special types of occupation and a high standard of management. The three factories from which data have been obtained all meet the requirements. I explained what I wanted to the superintendent or to some other responsible official in each case. He then selected the group or groups of operatives whom he thought proper, and placed the figures in my hands. There was no selection on my part, and in each case I have used all the figures, omitting only a few obvious errors amounting to perhaps a quarter of one per cent.

An investigation such as is here set forth may follow two modes of procedure. One is to take a few persons and investigate each minutely in order to eliminate all accidental variations. The other is to take many people and get rid of the personal variations by averages. The wages of a workman depend upon many factors aside from the weather. One man has been scolded by his wife because he did not earn enough last week, another wants to buy some clothes for his little boy, and a third was drunk last night. A sore toe may have far more influence than any possible climatic variation. To ferret out all these accidental circumstances is out of the question. Fortunately, they do not occur every day, and most people work weeks at a time without being much influenced by them. Moreover, when large numbers of people work in different cities and during different years, the individual circumstances neutralize one another. The day that

John Jenkins is disturbed because his boy has run away, Tony Albano is working hard because he is going to be married. Hence, by taking five hundred people we are able to eliminate accidental and individual circumstances and thus to reach a reliable result.

All three of the factories whence our data are obtained make hardware, but the work varies greatly. In one factory where Italians are the predominant nationality, brass sockets for electric lights, and other little brass fittings are made. One group of people was here engaged in tending machines. Some were turning out screws, others were putting pieces of sheet brass into automatic machines which turn out perforated plates. The work requires little skill, but much quickness and concentration. Another group, composed largely of Italians, was engaged in rolling and drawing hot brass, a heavy and somewhat difficult kind of work, requiring considerable strength. It is difficult because the brass must be used hot, and hence the men must work in abnormally high temperatures. At another factory, the one from which the largest number of records was obtained during three successive years, there were two main groups of men and two of women. The girls, from sixteen to twenty years of age, were Americans by birth, but of varied descent, being chiefly Irish, Germans, Scandinavians, English, and other north Europeans. Their work was the packing of hinges and screws, which are first wrapped in tissue paper and then placed in pasteboard boxes. This is a light, easy task in which dexterity and accuracy in picking up the right number of pieces are particularly important. For the first week or two when screws are packed, the tips of the fingers become sore, which makes the work proceed slowly. If a girl is changed from packing hinges to packing screws, her wages fall off for a time, but such changes are not frequent, and do not appreciably influence our figures. The men at this factory were of all ages, and were of the same races as the girls. They were engaged in grinding and buffing the hinges. The first operation is hard, heavy work. The hinges are held upon rapidly revolving emery wheels in order to grind them to a smooth surface. The other operation, buffing, is similar except that it is easier, for the hinges after being ground are polished upon rapidly revolving cloth buffs covered with emery dust. In the third factory, the operatives were of north European descent, almost all being native-born. Practically all, both girls and boys, were young, only a few being much over twenty years of age. The older girls leave to be married, and the boys, who are comparatively few in number, go elsewhere to find harder and hence better paid work. The work consists of the preparation of armatures and other wire coils for electrical purposes. Some operatives wind the wire upon rapidly revolving spools. Others put together the various parts of an armature. The work is light and not tiresome. It requires much dexterity and accuracy. Strings have to be tied at particular spots, pieces of paper must be inserted, the machines must be stopped when the right point has been reached, and little ends have to be grasped and inserted in their proper places. Taking our three factories together, the work ranges from the hardest to the lightest. It is of many kinds, requiring different degrees of strength and skill. The wages depend not only upon the amount of work completed, but upon the number of pieces rejected. In other words, the wages represent not only speed, but accuracy.

Let us now turn to the actual performance of the operatives. This is summed up in Figure 1. The four upper solid lines represent the work done week after week, each year from 1910 to 1913. In Figure 1 the work of only about 410 people has been used. The rest have been omitted because the figures are not complete for a whole year. In only one case has there been a deliberate omission of figures which cover an entire year. That was the Italians who draw hot brass and hence are subject to abnormal conditions of temperature. The method of procedure has been to find for each working day the average hourly wages for each group of operatives. Hourly wages have been used in-

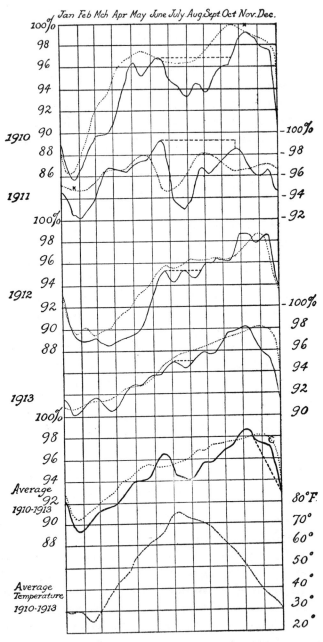

Figure 1. The Effect of the Seasons on Factory Operatives in
Connecticut (solid lines) and at Pittsburgh (dotted lines)

stead of daily, so as to make it possible to compare half-days
with whole. If part of the operatives were absent on any particular day, they were simply omitted, and the average for the
rest was taken. When the daily averages had been found, they
were averaged together by weeks. In doing this, a half-day, such
as the Saturdays in summer, was given only half as much weight
as a whole day, and days when part of the operatives were absent
or when the machinery was shut down for a while, were given a
correspondingly smaller weight. Thus allowance is everywhere
made for irregularities in the number of employees and the length
of time that they work. The final process consisted of combining
the different groups. In order that each individual may have the
same importance, all the figures have been reduced to percentages. In this way if a girl earned a maximum wage of twelve
cents an hour, it is called 100 per cent, while if a man's maximum wages were thirty cents, this sum also is called 100 per cent.
Thus the variations in the wages of the girl and the man have
the same weight in our final computations. Because of the enormous amount of work which would have been entailed, it was
not possible to reduce the wages of each individual to percentages, but only those of each group. Had it been possible to work
out each individual's wages separately, the results shown in our
curves would probably have been more striking than is now
the case.

In Figure 1 the height of the curves indicates the efficiency
of the operatives at various seasons for four successive years.
The fifth curve, heavier than the others, is the average of the
preceding four. Turning to the upper line, we see that in early
January, 1910, the efficiency of about 60 factory operatives in
Bridgeport was 88 per cent as much as during the week of
maximum efficiency that year. By the middle of the month it
had fallen to 86 per cent. Later it rose fairly steadily to 96
per cent at the end of April. Then it dropped a little, rose still
higher in June, and fell off distinctly during the summer, but

not so low as in winter. During the autumn it rose steadily until
early November, when it reached the highest point of the year,
after which it fell rapidly. In the same way each curve may be
traced week by week. I shall return to them shortly. Meanwhile,
it would be advantageous for the reader to look them over and
draw his own conclusions, picking out the features which are
common to all, and noting those which show different degrees
of intensity from year to year.

In Figure 1 it will be noticed that the solid lines never reach
100 per cent. This is partly because they have been smoothed,
and partly because they have been corrected to compensate for
the increased efficiency due to practice. The process of smoothing, as everyone knows, is used by mathematicians to eliminate
minor variations and thus permit the main trend of a curve to
be more apparent. It merely takes off the high points and the
low. The figures for three weeks are averaged, and the average
is used instead of the original figure for the middle week. In the
present case, and in practically all the curves in this book, the
process of smoothing has been performed twice on each curve.
If the letters a to e represent the average wages for five successive weeks, the figure actually used for the middle week, c,
is obtained from the following equation:

$$c = \frac{a + 2b + 3c + 2d + e}{9}$$

This process of smoothing can add nothing to a curve; it simply
takes away the less important details. If carried far enough it
would produce straight lines.

In addition to smoothing the curves I have corrected them
for the effects of practice. The curves for 1911, and 1912, and
1913 are all based on the same factory at New Britain. When
the wages for each year are averaged, we find that those for
1912 were 1.5 per cent higher than for 1911, and those for
1913 were 1.5 per cent higher than for 1912. This means that

constant practice caused the average employee, including both old hands and new, to be 1.5 per cent more skillful at the end of the year than at the beginning. Hence, from January onward, the curve rises a little until in December it is 1.5 per cent higher than it would be if the operatives had not grown more skillful. To eliminate this we simply tip the entire curve, raising the January end by three quarters of one per cent and depressing the December end by the same amount. The fluctuations, of course, remain unchanged.

In Figure 1, if there had been no correction, the highest and lowest points of the upper curve would lie at the points indicated by the crosses, and the other curves would be changed in corresponding ratios, there being no change at the end of June.

Turning to less technical matters, let us consider the degree of resemblance in the four upper solid lines of Figure 1. All are unmistakably low in January. Then from February to June we note a general rise, varied by minor fluctuations which differ from year to year. At the middle or end of June all reach a distinct maximum, although in 1912 and 1913 it is of slight proportions. Next we have a drop during the summer, pronounced in 1910 and 1911, but not at all prominent in 1912, and scarcely noticeable in 1913. Following this there comes a series of irregular fluctuations, differing from curve to curve, but in each case culminating in a strong maximum at the end of October or the beginning of November. Six weeks later, in the middle of December, another slight maximum is suggested, and then all the curves drop suddenly. In the average curve the minor fluctuations tend to disappear. They are more or less accidental, and represent peculiar conditions which pertain to one year but not to others. The features that have been named, however, show no sign of disappearing. They are five in number, namely, an extremely low place in midwinter, and a less pronounced low place in midsummer; a high point in June, a still higher point at the end of October, and a hump in mid-December. Much the most variable feature is the low place in summer. This is highly significant, as we shall shortly see.

Before we discuss the causes of the variability of the summers let us consider the meaning of the curves as a whole. In the first place, it is evident that, although details may vary from year to year, the general course of events is uniformly from low in the winter to high in the fall with a drop of more or less magnitude in summer. To what can this be due? Did the factories shut down in January, or run on part time, or decrease work because of lack of orders, or to overhaul the machinery and so forth? Do the high wages in October and November indicate a special rush of orders at that time? Of course, any variations in the way in which the factory is running must be reflected in the wages of the operatives, but in the present case this does not apply to the main variations, although it may apply to minor details. In neither of the two factories here considered were the responsible heads able to offer any explanation of the peculiarities of the curves on the basis of factory management or the exigencies of business. Both are engaged in making staple articles, the chief demand for which comes in the spring when building operations begin. There is no Christmas rush on hinges and electric light sockets. After Christmas the factories shut down for a few days at the beginning of the year, but that ought to increase rather than diminish the hourly earnings. When operatives are working only part time they feel the need of earning as much as possible each hour. If part of the hands are laid off, that would also increase the average hourly wages, for the weaker ones would be dropped, and the average ability of those who remain would be high.

In this connection, it is important to understand that in these factories a man is free to work as hard as he wishes at any time of the year. The managers have deliberately adopted the policy of getting as much work as possible out of each operative. Overhead charges for interest, superintendence, bookkeep-

ing, salesmen, and other outside expenses, and also the charges
for unproductive labor such as engineers, janitors, and the like
are no greater no matter how hard the productive employees
work. If the producing operatives should double their output,
most of the other expenses would scarcely increase at all. Hence,
it would not only be possible to pay double wages for double
work, but it would be profitable to the factory even if it paid
perhaps $2.50 where now it pays $1.00. In view of these conditions, both factories have adopted systems whose special object
is to encourage extra exertion. In one case, part of the men work
upon what is known as the "premium" plan. The management
and the men have agreed that the various tasks shall be rated
according to the number of hours which they may fairly be supposed to require. If a man performs an eight-hour task, he is to
be paid for eight hours' work, no matter whether he does it in
six hours or ten. If, however, he finishes the work in less than the
stipulated time, he goes to work at another task for the rest of
the period. For half of this time he is to be paid, while the factory
gets the benefit of the other half. For example, if an eight-hour
task is finished in six, the operative works two more hours. He
is then paid for nine hours although he has only worked eight,
while the factory gets ten hours' work and pays for nine. Thus
both are the gainers. In one case the managers made a mistake
in deciding upon the number of hours needed for a certain task.
It had never been done quickly, and no one knew how rapidly it
might be done. The man who does it soon earned ten or twelve
dollars a day, where he formerly earned perhaps two and a half
or three. Inasmuch as the management had agreed not to change
the rates, they stuck to their bargain. The task only occupies
one day each month, and the matter is not serious. Moreover,
even though the operative earns such high wages, the work
actually costs the factory less than when he was earning two
dollars and a half.

In the other factory the girls are stimulated by bonuses. That is, they are not only paid for their work, but if they do more than is expected they are paid a bonus. For example, if a girl's wages averaged about a dollar a day, and she did work worth $1.20, she did not receive $1.20, but $1.25 or even $1.40. The factory finds this worth while because so much more can be produced without any increase in charges for interest, office work, and other overhead expenses. When this bonus system was first introduced, it produced only a slight effect. The girls did not seem to care about the bonuses and made little effort to get them. Then the management realized that the parents were getting the extra money, and so it made no difference to the girls, most of whom gave their pay envelopes unopened to their fathers or mothers. Thereafter the bonus was not put in the pay envelope, but was handed out in loose change. The girls kept it and began to work hard. In the third factory, whose figures are not extensive enough to be used in Figure 1, but which enter into other computations, a similar system is employed. A limit is set for each task. If the work is performed within that time a bonus is paid. Otherwise the operatives receive only the regular pay, no matter how much time they spend. The introduction of this system has increased the output of the factory enormously. Inasmuch as the various systems of bonuses and premiums are equally applicable at all times of the year, it seems impossible to find in the factories themselves any reason why earnings should be very low in January, moderately low in July, high in June, and very high in November.

We seem forced to search outside of the factories for the reasons for our seasonal fluctuations of wages. Such things as panics, hard times, or strikes would certainly cause a general change in the conditions of work. but nothing of the kind occurred during the period under consideration. Moreover, such events do not recur at the same time each year. Aside from the seasons, the only event which recurs regularly year after year at the same time and which is important enough to cause variations in wages is Christmas. Its effect can be seen unmistakably in each of the solid year-curves. In that for 1910 it appears in the little hump which culminates during the next to the last week in December. In the other three it comes a week earlier because this factory does not pay the week's wages on the Saturday of the week in question, but a week later, after there has been time to check up the work and make allowances for that which is poorly done. Hence, money for Christmas must be earned before the middle of December. If there were no such thing as Christmas, the wages would probably drop off in the way shown by the dash line in the average curve of Figure 1. After Christmas the wages probably drop somewhat lower than would otherwise be the case, for there must be a reaction from the previous effort, but it is noticeable that the wages do not reach their lowest ebb directly after Christmas, but keep on falling for nearly a month. Something else keeps them low. The Christmas hump is significant chiefly because it shows unmistakably that an outside stimulus which applies to all the operatives produces a distinct result. We may properly infer that the other permanent features of our curves are also due to some outside force which influences all the operatives. That force must be connected with the seasons, and it must be far more powerful than Christmas, for its effects are far greater. There seems to be no recourse except to ascribe the fluctuations of the curves to climate.

The verity of the conclusion just reached is strongly confirmed by comparison with other regions and other types of human activity. Figure 2, which, for convenience, is here divided into two overlapping portions, presents a series of curves arranged according to climate, those from regions with cold winters and cool summers being at the top, and cool winters and hot summers at the bottom. The curves range from the Adirondacks in northern New York to Tampa in southern Florida, and include one from Denmark. With them I have repeated some of

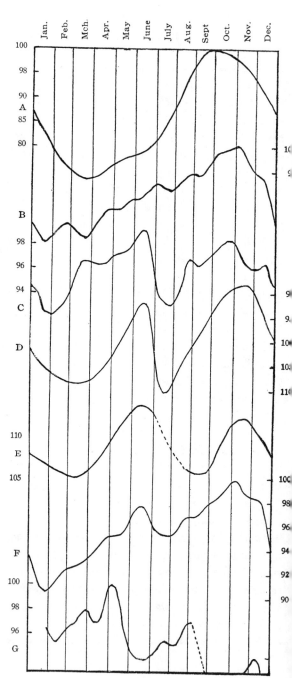

Figure 2a. Human
Activity and the
Seasons

A. Gain in Weight of
1200 Tubercular
Patients at Sara-
nac Lake, N. Y.,
1893-1902.

B. Work of 160 Fac-
tory Operatives in
Connecticut, 1913.
Repeated from
Figure 1.

C. Work of 60 Fac-
tory Operatives in
Connecticut, 1911.
Repeated from
Figure 2.

D. Deaths in the
State of New York,
1892-1906. In-
verted. In the
scale of this curve,
100 represents the
average death
rate.

E. Strength of 90
School Children in
Denmark, 1904-6.

F. Work of 410 Fac-
tory Operatives in
Connecticut, 1910-
1913. Repeated
from Figure 1.

G. Work of 65 Girls at
Winston-Salem, N.
C., 1914.

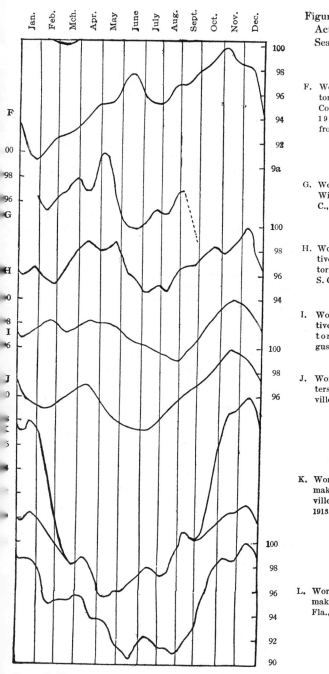

Figure 2b. Human Activity and the Seasons

F. Work of 410 Factory Operatives in Connecticut, 1910-1913. Repeated from Figure 1.

G. Work of 65 Girls at Winston-Salem, N. C., 1914.

H. Work of 120 Operatives in Cotton Factories at Columbia, S. C., 1912-14.

I. Work of 120 Operatives in Cotton Factories near Augusta, Ga., 1912-14.

J. Work of 57 Carpenters at Jacksonville, Fla., 1911-14.

K. Work of 400 Cigarmakers at Jacksonville, Fla., 1911, 1913, 1914.

L. Work of 2300 Cigarmakers at Tampa, Fla., 1912-14.

NOTE. In Figures 2a and 2b the unit is a year's work of an individual. Thus "120 Operatives, 1912-14," means an average of 40 per year for three years.

the curves of Figure 1 for the sake of comparison. The most remarkable feature of this series is that, although there is great diversity of place and of activity, all the curves harmonize with what would be expected on the basis of Figure 1.

The first curve, A, is based on the work of Lawrason Brown, a physician who has published records of the weight gained by patients suffering from pulmonary tuberculosis at a sanatorium at Saranac Lake in the Adirondacks. A gain in weight in this disease is a favorable symptom, for one of the most marked effects of tuberculosis is to cause a wasting away of the flesh. In the present tabulation the patients who lost weight are not included, and a drop in the curve does not indicate loss of weight but merely a decreased rate of gain. If the patients who lost weight were also included, however, the form of the curve would still be the same, according to Brown. The Adirondacks, as everyone knows, have long cold winters, while the summers are delightfully bracing, being warm enough to be pleasant, but never hot enough to be debilitating. Hence, from about the first of April to the end of September the sick people make a marked gain. During the other six months, although they may gain more than would be the case in their own homes, they do not find the climate nearly so advantageous as in summer, and the disadvantage increases until the snow disappears. The next curve, B, is a repetition of the Connecticut curve for 1913. That year the winter was by no means so severe as is ordinarily the case in the Adirondacks. Hence, the curve does not remain low quite so long as does A, and does not begin to fall so soon. The summer, however, was almost as cool as among the Adirondacks, and hence there is no drop during July.

The next pair of curves represents a year with a hot summer in Connecticut, C, and the death rate for fifteen years in the state of New York, D. The curve for deaths has been turned upside down, so that high places represent few deaths, that is, high vitality corresponding to high energy in the factory operatives. In New York State as a whole the effect of the summers is very different from what it is in the Adirondacks. The cities swelter for a few weeks in July, and that sends the death rate up enormously, especially among children, who are quickly taken sick, and who either die after a few days' illness, or recover. That is why the curve drops so sharply in mid-summer. In the winter, on the contrary, although it drops almost equally low, the maximum number of deaths per day does not come till March, although by that time the average energy of operatives has risen considerably. This is because people become sick in January and February, especially those who are elderly, and finally die after lingering illnesses quite unlike those of children.

The death rate of other places might be used quite as well as that of New York. The Japanese rate, for instance, is as follows, the figures being those for the ten years beginning with 1899. The figures represent percentages of the normal. Those for the state of New York, computed on the same basis, are added in parentheses:

January,	104 (105)	May,	85 (100)	September,	118	(97)
February,	108 (108)	June,	84 (90)	October,	102	(89)
March,	103 (109)	July,	97 (110)	November,	96	(86)
April,	90 (106)	August,	116 (104)	December,	98	(93)

Here the course of events is almost the same as in New York, but with significant differences which harmonize with the climates of the two places. Winter in Japan is less severe than in New York, and its effects do not last so long, for the highest mortality is in February instead of March. The Japanese summers, on the contrary, are characterized by prolonged heat, and also by great humidity, especially during the rainy season from July to September. At the end of this period the mortality is at a maximum. The debilitating effect of the summer lasts so long that November and December have a higher death rate than May and June. The late spring is especially favorable, not only because of its own excellent character, but because it follows a winter which is not severe enough to be highly disadvantageous.

Curves E and F represent the strength of ninety school children in Copenhagen as measured by Lehmann and Pedersen, and the average energy of factory operatives for four years in Connecticut. The Danish measurements were carried on during the school years of 1904-1905 when sixty children were tested weekly, and 1905-1906 when ten were tested daily. By combining the two years into one and making allowance for the fact that children grow stronger from month to month just as factory operatives grow more skillful, we obtain curve E in Figure 2. Since neither summer nor winter is especially severe in Denmark the dip at the two seasons is the same. The maxima in June and November are almost synchronous with those in Connecticut. The minima are both delayed six or seven weeks, but the winter minimum in March agrees with the maximum death rate in New York. The summer minimum ought possibly to come in July or August, but the figures for those months are not obtainable, for during that time the schools in Copenhagen have vacation. In addition to this we should expect the Danish curve to lag a little behind that of Connecticut because of the maritime climate. Inasmuch as Denmark is constantly swept by west winds from the ocean it does not so quickly grow cool in winter nor warm in summer as does Connecticut, where the prevailing winds are from the continental interior, which of course becomes rapidly warm in summer and cool in winter. Thus it appears that the strength of Danish children and the energy of factory operatives in Connecticut have an almost identical relation to seasonal variations of climate.

Judging by curves C to F in Figure 2 one might hazard the hypothesis that man is subject to a seasonal rhythm which repeats itself wherever he goes without regard to the climate. On this basis one would expect maxima of efficiency in June and November in all parts of the world. In curves A and B, however, we have already seen that where the summers are particularly favorable and the winters unfavorable this rhythm breaks down, and the June maximum and summer minimum disappear. If we go farther south to places where the winters are favorable and the summers very hot, we find a change in the opposite direction, for the winter minimum tends to disappear, and the summer minimum greatly increases and shoves the two maxima more and more into the winter until the two coalesce. This is evident in curves G to L. These represent variations in the wages of piece-workers in southern factories, compiled according to the method used in Connecticut. Curve G shows the work of sixty-five Anglo-Saxon girls in a tobacco factory in Winston-Salem, North Carolina. They were pasting labels on cans. Notice how their winter minimum comes in early May instead of June. In September the curve drops suddenly. This is because at that time the effect of the war began to be felt, the price of cotton fell so low that the South was in great distress, and the sale of the goods made by this factory began to be curtailed. Therefore the girls were not given as much work as they could do.

Curves H and I are from cotton mills in South Carolina and Georgia, and each represents two mills. In South Carolina the two mills are close together at Columbia, while in the other case they are fifteen or twenty miles apart, one being in Georgia near Augusta, and the other across the Savannah River in South Carolina. The operatives in all cases are of pure Anglo-Saxon stock, chiefly of the "poor white" class. Men and women are included in nearly equal numbers. Part are weavers, while others, engaged in the occupations known as slubbing, spooling, and speeding, tend machines which spin the thread and wind it on bobbins ready for the weavers. In all cotton factories the air in the weaving room, and to a less extent in the others, is kept at a high temperature and a high humidity. This is necessary because when the air becomes cool, or especially when it becomes dry, the thread is apt to break and cause blemishes in the cloth.

Hence, in factories where high-grade goods are manufactured the inside temperature is so abnormal and the amount of goods produced depends so largely on the breakage that it is almost impossible to obtain satisfactory figures. In the factories here considered, however, nothing but coarse cloth is manufactured. The breaking of the thread does little harm, and relatively slight attention is paid to the temperature and humidity of the weaving rooms. Moreover, for slubbing, speeding, and spooling, the temperature and humidity make far less difference than for weaving. Hence, the variations in the amount of goods produced per person depend largely on the energy of the operatives in watching their machines and preventing them from standing idle because of broken threads, empty bobbins, or other accidents. The exigencies of business, that is, the demands for goods, make no difference to the operatives so far as their production per hour is concerned, for the machines run at a uniform speed whether the factory runs one day a week or six. The cotton mill curves are essentially the same as that of the tobacco factory. In H there is a double spring maximum, due to accidental circumstances, but the true maximum would probably come about the end of April. In I the spring maximum comes still earlier, that is in mid-April, as is appropriate to a place so far south. The autumn maxima, on the other hand, come later than in Connecticut, one being in early December and the other toward the end of November.

The work of carpenters in Jacksonville, as shown in curve J, is different from anything else that is here considered because it is performed out of doors. The fifteen men per year whose records are here used were engaged in making the same kind of repairs time after time. A careful record of the hours that they spend is kept, but the number varies greatly on account of the weather. If it rains they cannot work. Summer is the rainiest period, but that does not tend to diminish the amount of work done per hour. In fact it increases it. The rain comes in hard showers, and while it is falling the men rarely try to work, and the time is not reckoned. When the rain is over they work better than before because the air is cooler, although still far from being cool. In winter, on the contrary, from December to March, the rain is a pronounced hindrance. It often comes in the form of a drizzle, and the carpenters try to keep on working while it is falling. Moreover, after the rain the wood is wet, there is apt to be a chilly wind, the hands feel numb, and everything is opposed to great efficiency. Yet in spite of this, more work per hour is done in February, the worst winter month, than in May, June, July, or August. If these men were at work in well-protected sheds which were heated on the occasional cool days, there is little doubt that in December their curve would reach a maximum higher than that now reached in November, while even if the following months were not still better, they would at least show no pronounced drop.

The lower two curves, K and L, represent the work of cigar-makers at Jacksonville in northern Florida and Tampa in the southern part of the state. Those in Jacksonville were mostly Cubans, nearly two thirds being negroes, and the rest of Spanish descent. At Tampa only a handful of negroes is included, but a large sprinkling of real Spaniards is found among the Spanish Cubans. The curves for the cigar factories are compiled on a different basis from the others. The reason is that there are no definite hours. The factories are open twelve hours a day, usually from 6 a.m. to 6 p.m. The operatives saunter in as they please, provided they do not come later than 8 a.m., and leave when they choose, although an attempt is made to let no one depart before 4 p.m. While at work they sit close together at tables, and talk volubly except when a hired reader is vociferating the news from a Spanish newspaper. At some time in the morning they go out for a lunch, but are rarely gone as much as half an hour. Otherwise they stay at their work till it is finished.

Since there are no fixed hours, we cannot measure the exact

earnings per hour, as we have done in other cases, but only the earnings in proportion to the time that a man might have worked if he had chosen to do so. In other words, we measure partly the actual capacity for work, and partly the inclination to work. In general the two seem to vary together, but the work of the New York State Commission on School Ventilation has shown that during short periods of high temperature the capacity may remain unimpaired, while the inclination declines. In the practical work of life a lack of inclination is almost worse than a lack of capacity.

During the warmer half of the year the possible working time in the Florida cigar factories may be properly reckoned as eleven and a half hours. In winter, however, the light at morning and evening is not adequate for the somewhat exacting work of cigar-making. Therefore the men are not allowed to begin so early as in summer, nor to work so late. The exact time depends on the degree of cloudiness as well as the height of the sun. The factory managers say that in December the working time is curtailed an hour and a quarter or more for the month as a whole. In order not to make the winter production appear unduly large, I have reckoned that during the shortest week—not month— the working time is an hour and nine minutes, that is, 10 per cent less than in summer. Before and after that date it steadily increases to the solstices, when it reaches the normal. Thus we get the lower curve for Tampa. It drops low in summer and rises to a single maximum in winter. At Jacksonville the variations in the length of the working day on account of light are less than at Tampa because a lower grade of cigars is made, and hence the men are allowed to work under less favorable conditions of light. Inasmuch as the exact effort of dark mornings and evenings cannot be determined, I have drawn two lines at each end of the curve. The lower shows the wages if no allowance is made for light, and the upper if the full Tampa allowance is made. The actual truth lies between the two. For our present purpose this uncertainty makes no difference, since in either case we have the summer minimum and winter maximum which all our other studies would lead us to expect in this latitude.

The exigencies of business have more effect on the work of the cigar-makers than on that of the other operatives employed in Figure 2, but they do not determine the main fluctuations of the curves here used. In some cigar factories, to be sure, if business is slack the employees are often not allowed to make more than half or two thirds the usual number of cigars. For this reason I have omitted two factories whose figures I worked up, but whose curves I finally found to be almost wholly controlled by the supply and demand of the business. In the three factories which were finally used, however, that is, one at Jacksonville and two at Tampa, the operatives are only rarely placed on a limit. It is too expensive, especially where high-priced cigars are made, for four cigars a day have to be allowed to each man for "smokes." Each man smokes his full number, if not more, no matter whether he makes one hundred cigars or two hundred. The rush season for cigars begins in June or July and becomes increasingly intense until about the middle of November, by which time most of the Christmas orders have been received. Business is dullest in January and February. The operatives, however, know nothing about this, except as they see that men are taken on or discharged. The frequency of changes in the number of employees makes the cigar-maker's life hard, and accounts for much of his proverbial shiftlessness.

Another thing which affects the wages of cigar-makers is the dampness of the air. During the warm, damp days so characteristic of the Florida summer, the tobacco is very pliable and easily worked, while on dry, winter days its brittleness causes it to break so that the work is hampered. If it were not for this the difference between summer and winter would be intensified.

The most striking proof of the effect of the seasons is yet to be recorded. It consists of a series of data corresponding to those of the Connecticut factories, but based on the work of operatives in a large factory engaged in making electrical apparatus at Pittsburgh, Pennsylvania. The employees whose wages were investigated were employed in winding wire coils, assembling the parts of motors, and other similar operations which demand accuracy and speed. The admirable way in which the records of this company are kept renders the figures of great value, but lack of time and funds has made it necessary to limit the present investigation to monthly, or, in 1912, bi-weekly averages of hourly earnings. For this reason the resulting curves, which have been inserted as fine dotted lines in Figure 1 (page 59), are smoother than those of Connecticut where the daily earnings have been utilized. The number of piece-workers on which these Pittsburgh curves are based is shown in the following table:

1910. Approximately 950 men and girls in winding section.
1911. Approximately 750 men and girls in winding section.
1912. Twenty-seven girls, winders; 42 men, tinners, blacksmiths, painters. In this case all the operatives were especially steady hands who worked throughout the year. In the few cases where they were absent, interpolation has been resorted to. Hence this year's curve is more reliable than the others which are based on all the operatives in a given section or in the whole factory without regard to whether they worked steadily.
1913. Approximately 7000 men and girls in the entire factory.

The general form of the curves for Pittsburgh and Connecticut is obviously the same. In 1910 notice the deep dip in January, and the moderate drop in summer. The next year, 1911, presents quite a different aspect. Because of the hot summer, the depressions in January and July are almost equally deep, the difference between the highest and lowest points is less than in most years, and the autumn maximum does not rise high above that of May or June, as is usually the case. The curves for 1912 both show a deep depression in winter which lasts unusually long. During the summer, on the contrary, there is not so great a decrease in efficiency as during the previous two years. Finally, in 1913, both curves rise almost steadily from midwinter to late fall, with only a slight drop in summer.

The agreement between the curves for Connecticut and Pennsylvania is far too close to be accidental. At Pittsburgh, just as at the other factories, variations in the total number of employees form an accurate measure of the demand for work, but these by no means vary in harmony with the actual production per operative. Often the average amount of work done by a given group of individuals, or by all the piece-workers, declines when the number of operatives increases, but quite as often the reverse is true. Hence the conditions under which the factories are run do not explain the variations in wages. Moreover, it stands to reason that the same irregular variations would not occur season after season in an electric factory in Pittsburgh and in brass and hinge factories in Connecticut 400 miles away unless all were under the same control. The only common controlling factor which varies in harmony with the curves of Figure 1 is the general character of the seasons. This is essentially the same in both places.

We have now seen that from New England to Florida physical strength and health vary in accordance with the seasons. Extremes seem to produce the same effect everywhere. The next question is whether mental activity varies in the same way. Lehmann and Pedersen made a series of tests of the ability of school children in addition. Their general conclusion is that mental work varies in the same way as physical, but reaches its highest efficiency at a lower temperature. This agrees with the investigations of a few other scientists, and with the general conclusions of the world as summed up in the old adage, "No one is

worth a tinker's dam on whom the snow does not fall." Before
we can accept this, however, tests are needed on a large scale.
The most feasible method at present seems to be by means of the
marks of students in such schools as West Point and Annapolis.
There the young men live an extremely regular life with a minimum of outside distractions. Their recitations are graded with
great severity and regularity, and a given subject is often
taught six days in the week. The marks are handed to the heads
of departments at frequent intervals and are posted where the
students can see them. No class is taught in divisions of more
more than ten or twelve, so that every student has a full opportunity to show how well he is prepared. In order to avoid all
chance of favoritism the instructors do not keep the same division month after month, but change every few weeks. Altogether
it would be hard to devise a system which more thoroughly
eliminates the human and accidental factors. As an instructor at
West Point put it: "We are not really teachers. We are just
put here as officers to see whether the cadets have studied their
books, and to decide how many marks to take off." This is pre-
eminently true in mathematics, where the solution of a problem
is either right or wrong and can be marked accordingly.

When I broached my plan to the superintendents at the two
academies, it was received with much interest, and every facility
was placed at my disposal. I take this opportunity to express
my warm appreciation of their courtesy. Some of the instructors
were commissioned to see that the proper records were available.
The marks of individuals were, of course, not necessary. The
various marks for each day or week were merely added, and
averaged. The data here employed embrace the following: (1)
The weekly averages in mathematics for the first-year or entering class at Annapolis for the six academic years beginning with
1907-1908 and ending with 1912-1913. These classes recite six
times a week. (2) The daily marks for the first-year class in
English at Annapolis for the year 1912-1913. This class recites

four times a week. (3) The daily marks in mathematics for a year and a half for the classes entering West Point in 1909 and 1910. Recitations are held six days a week. The classes at Annapolis average about 220 in number and those at West Point about 120. The entire number of students whose marks have been used is between seventeen and eighteen hundred, but as some of the marks cover a period of a year and a half, the total is equivalent to about nineteen hundred students for a single year.

All these marks have been combined into the three lower curves of Figure 3. Before discussing them a few words should

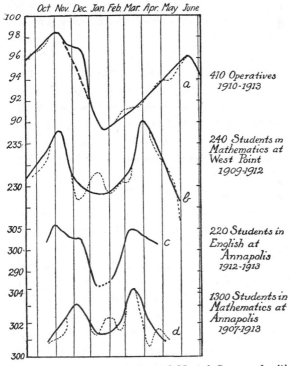

Figure 3.   Seasonal Variations of Mental Compared with
Physical Activity

be said as to the method of preparation. The systems of marking at the two academies are quite different. At Annapolis the department of mathematics tries to keep the average as nearly uniform as possible. If the instructors discover that the average is rising or falling they mark more severely or leniently to counteract it. At West Point, on the other hand, the marks regularly begin high at the opening of the term and fall steadily toward the end. There is no attempt to keep them at a uniform level, but the instructors merely mark harder and harder or give more and more work as time goes on. Both systems tend to mask the effect of the seasons. The influence of the deliberate attempt to keep the marks at a uniform level at Annapolis is largely overcome by using a series of six years. The irregularities of one year counteract those of another except where special circumstances such as vacations interpose a disturbing element at the same time each year. In the English department at Annapolis there is less stringency about keeping the marks at a uniform level, and those of a single year show clearly the normal seasonal trend. At the end of the year, however, I have omitted the two weeks before examinations because there was then a sudden spurt accompanied by abnormally high marks. Otherwise all the Annapolis marks without exception have been employed in computing the curves of Figure 3.

At West Point it has been necessary to eliminate the effect of the steady fall. The method is the same as in the correction for increasing practice. In order to eliminate the effect of such things as football games, holidays, examinations, reprimands, or other circumstances which clearly have nothing to do with climate, I have omitted all the days whose marks fall more than 10 per cent above or below what would be expected at that particular date. Omissions of this sort are such a common procedure in astronomical and physical measurements that the mathematician requires nothing more than a mere mention of what has been done. To the layman it may seem that they are of great importance. In reality they rarely alter the general form of the final curves, for exceptionally high figures balance exceptionally low. In the second curve of Figure 3 the effect is slight except upon the first weeks in January. There the minor maximum which occurs just after the Christmas recess is only about half as large as it would be if no data were omitted. At Annapolis it is not necessary to omit the days of special events because the marks are not subject to such wide fluctuations. It is interesting to notice that the classes in mathematics there are influenced by the vacation, which comes at the end of January, just as at West Point. The English marks, on the contrary, are uninfluenced, probably because English is an easier subject than mathematics. Moreover, as it is taught fewer days per week, and hence has less weight in determining the final marks for the work of the whole year, the students do not devote so much energy to it.

By this time the reader has doubtless interpreted Figure 3 for himself. The upper line is the standard average curve for factory operatives in Connecticut. It is the same as the average curve of Figure 1, except that it begins in September instead of January. It is placed here to permit a comparison of the physical work with mental. The curves of mental activity all resemble it in having two main maxima, in fall and spring. At West Point, where the climate is essentially the same as in Connecticut, the mental maximum in the fall comes about ten days later than the factory maximum, while the spring maximum comes two and a half months earlier. Both occur when the mean temperature is a little above 40° F. At Annapolis the maxima are, as it were, pressed toward the winter. The fall maximum in English, to be sure, begins early in November, but lasts till the middle of December. Since it represents the work of only a single year, it is less important than the curve of mathematics, whose fall maximum does not come till the first half of December. The spring maxima of both curves come in the middle of March. At Annapolis, just as at West Point, the time of best work is when the mean temperature is not far from forty degrees.

Summing up the matter, we find that the results of investigations in Denmark, Japan, Connecticut, Pennsylvania, New York, Maryland, the Carolinas, Georgia, and Florida are in harmony. They all show that except in Florida neither the winter nor the summer is the most favorable season. Both physical and mental activity reach pronounced maxima in the spring and fall, with minima in midwinter and midsummer. The consistency of our results is of great importance. It leads to the belief that in all parts of the world the climate is exercising an influence which can readily be measured, and can be subjected to statistical analysis. It justifies us in going on with confidence to ascertain exactly what effect is produced by each of the climatic elements, such as temperature, humidity, and pressure.

CHAPTER V

THE EFFECT OF HUMIDITY AND TEMPERATURE

HAVING seen that both physical and mental energy vary from season to season according to well-defined laws, let us now investigate the special features of seasonal change which are most effective. Temperature is far the most important, but before considering it, let us discuss those of minor importance. One of these is light. Many students have ascribed great influence to sunlight, and to its variations from season to season, or from one part of the world to another. For example, C. W. Woodruff, an army surgeon, has written an interesting book on *The Effect of Tropical Light on White Men.* Its main thesis is that the backwardness of tropical countries is due to excessive sunlight. The actinic rays at the blue end of the spectrum, especially those beyond the limits of vision, possess great chemical power, as is evident from the fact that by their aid photographs can be taken even when no light is visible to the naked eye. Such rays, when they fall upon the human body, are thought to stimulate the cells to greater activity. At first this is beneficial: if it goes to excess the cells apparently break down. The process is analogous to the ripening of fruit. A moderate change in the green tissues produces the highly favorable condition of ripeness: more brings on decay. Thus while the return of the light after the winter of the temperate zone may be beneficial, excessive light may be highly injurious.

So far as our factory operatives are concerned, no effect of light is to be discerned in the South, while in Connecticut it is at best only slight. The heavy line next to the bottom in Figure

1 (page 84) shows that from mid-September to the middle of November the amount of work increases, although the days are growing shorter. This is exactly opposite to what would be expected if the shortness of the days were of primary importance. Moreover, in June when the days are longest we find a sudden drop. If the length of the days had much to do with the matter, there is no reason why more work should be done in November than in June. Nor should we find that a shortening of the days during September is accompanied by the same kind of increase in efficiency which is seen in March when the days, although of the same length as in September, are growing longer instead of shorter. For all these reasons we assign only slight importance to variations in the amount of light. Nevertheless, some effect can apparently be detected. Compare the two lower curves of Figure 1. In spite of the low efficiency occasioned by the winter's cold, the curve of work begins to rise sooner than does the curve of temperature which is placed below it. The first appreciable lengthening of the days in January may cause this by its cheering and stimulating influence.

The line of reasoning applied to light applies also to the possibility that the variations of the curve of work depend on the extent to which people are shut up in the house. Obviously, this has nothing to do with the two maxima in November and May, nor with the minimum in July. In November people's houses have been shut up for a month, more or less, while in May and July they are wide open, or at least as wide open as they ever are. The extremely low minimum in January, however, is probably due at least in part to the necessity of shutting up the house in winter. In October the weather becomes so cold that people begin to shut up their houses; they live in stuffy, unventilated quarters, and fail to take exercise in the open air. By the middle of November this has had time to produce an effect which naturally becomes more and more marked as the weeks go on. This would harmonize with the decline of energy from November to the middle of January. In January, however, the decline ought not to cease if it is due chiefly to confinement within the house. It ought to continue until about the middle of March, for not till that time do people in Connecticut begin to let in the outside air, and not even then to any great degree. As the curve of work has risen distinctly by that time, some other factor must intervene, presumably the increase of light to a slight extent, and the rise of the temperature to a larger extent.

A third factor to be considered at this point is the relative humidity of the atmosphere. A sharp distinction must be drawn between the humidity of the outside air and that which prevails within doors. Physicians, students of factory management, school superintendents, and many other people have repeatedly discussed the supposed harmful effects of the dry air in our buildings during the winter. A much more fully attested fact is the harmful influence of great humidity during hot weather. We are more conscious of this than of the harm arising from excessive dryness. This does not necessarily mean that the total effect is worse than that of dryness, however, for hot, humid days are much rarer than the winter days when the air in our houses is drier than that of the majority of deserts.

So far as our curves of work are concerned, humidity does not seem to be responsible for the fluctuations except as it is influenced by temperature. In other words, the average humidity of the outside air from season to season does not vary in such a way as to cause maxima in May and November, and minima in January and July. The average humidity of the outside air in November and in January is not greatly different. Nevertheless, the *inside* humidity is probably an important factor in causing the low efficiency of midwinter.

The relation of work and humidity among the factory operatives of Connecticut is illustrated in Figure 4. There the year has been divided in three parts: (1) winter, (2) spring and autumn, and (3) summer. In each part all the days having a given humidity have been averaged together, and the smoothed results have been plotted. The heavy, solid lines represent what I believe to be the true conditions when other disturbing elements are removed, while the dotted lines show the actual figures. In winter the dampest days are unmistakably the times of greatest efficiency. We may shiver when the air is raw, but we work well.

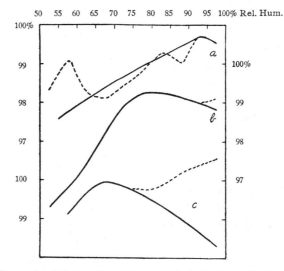

Figure 4. Relative Humidity and Work in Connecticut

a. Winter. b. Spring and Fall. c. Summer.

This is partly because in winter the dampest third of the Decembers, Januaries, etc., averages nearly 2° F. warmer than the driest third. Moreover, moist air at any given temperature feels warmer than dry and hence is less likely to cause people to overheat their houses.

In the spring and fall, when the temperature ranges from freezing to 70°, with an average of about 50° F., the best work is performed with a relative humidity of about 75 per cent. In other words, neither the dry nor the wet days are the best. The summer curve is the most complex of the three. It rises first to a maximum at 60 or 65 per cent, then falls, and once more rises to a higher maximum. The first maximum seems to be due to humidity, the second to temperature. A hot, damp day is unquestionably debilitating. The majority of the dampest days in summer, however, are comparatively cool, for they accompany storms. The coolness counterbalances the humidity, and people's efficiency increases. Hence, we disregard the right-hand maximum and conclude that with an average temperature of 65° to 70° a relative humidity of about 60 per cent is desirable.

The most unmistakable feature of the curves as a whole is that they show a diminution of work in very dry weather. This presumably has a bearing on the low level of the curve of energy in winter. At that season the air in our houses ought to have a humidity of 60 or 65 per cent, but most of the time the figure is only 20 or 30. On very cold days the percentage is still lower. For instance, if the outside air has a temperature of 14° F. (−10° C.) and contains all the moisture it can hold, which is usually not the case, its relative humidity when it is warmed to 70° F. will be only 12 per cent. Even on days when the outside humidity rises to 100 per cent and the temperature is 40°, the air in an ordinary steam-heated house has a relative humidity of only 35 per cent, which is far below the optimum. Apparently, this extreme aridity is debilitating. It probably dries up the mucous membranes in such a way as to increase our susceptibility to colds. In this way it may be an important factor in causing February and March to have the highest death rate of the year. There has been a good deal of discussion as to the actual importance of atmospheric humidity, and no small amount of disagreement. We shall return to the matter later when we study health.

While the effects of light, of closed houses, and of excessive dryness explain part of the fluctuations of the curve of work, they have little bearing on any season except the winter. An-

other matter which may be suggested in this connection is vacations. These, like many other conditions of human life, are
largely seasonal. Do people work fast in the fall because they
have been rested by vacations? In professional occupations and
in business this certainly seems to be the case, but not among
factory operatives. As a rule such people do not take summer
vacations. They usually stop work at irregular intervals, or
else after Christmas when many factories shut down or work on
part time for a few days to prepare for the new year. The form
of our main curve, however, shows that neither at this time nor
in summer do vacations produce any appreciable stimulating
results. If they were the cause of fast work, the curve ought to
be the highest within a few weeks after the people return to work,
but this is not the case. During the vacation period of July and
August the amount of work is moderately low, and in early
January, after the Christmas break, very low. At the end of
August it begins to increase, and increases steadily for two and
a half months. The maximum in November is so long after the
vacation period that it can hardly have anything to do with it.

What has just been said has an important practical application. There is a common idea that people need vacations in
summer. Of course there are strong arguments for this, since
pleasant recreation is then possible out of doors. Nevertheless,
the need is apparently greater in winter than in summer. To
meet this it is probably wise that work should be light during
the winter. Already, as everyone knows, many factories run on
part time during the first few weeks of the year, and now we see
that there are strong physical reasons for this. Another important suggestion afforded by our curves is this: If the operatives
of a factory, or people engaged in any other kind of work, are
to be speeded up, the time to do it is when nature lends her aid.
To speed up at the end of January is analogous to taking a
tired horse and expecting him to win a race. Later in the year,
however, during the spring, especially in May, people may ap-

parently be pushed to the limit, and will not suffer, because their energies are naturally increasing. This is still more the case in October and early November. After the middle of November pressure may produce important results, as we see at Christmas. Nevertheless, the chances are that if continued it will produce undue exhaustion, followed by a serious reaction. Possibly the nervousness of Americans is due partly to the fact that although we relax somewhat in summer, we keep ourselves at high pressure through the winter when the need of relaxation is greatest.

Turning now to temperature, we see that in Figure 1 (page 84) the lower curve, showing the march of temperature through the year, and the Connecticut curve just above it are similar in many ways. Both are low in midwinter. From February onward they rise together until about the middle of June. Then the efficiency curve falls while the other goes on rising, a condition which fully accords with ordinary experience. The fall of the efficiency curve begins when the average temperature has risen to about 68°. When the temperature stops rising, the work stops falling, and then remains nearly steady through July. At the end of July the mean temperature has fallen to about 71°. During the succeeding period of favorable temperature the two curves disagree, for the amount of work goes up while the temperature falls. When the average temperature falls below 48°, however, and begins apparently to be unfavorable to physical exertion, the curve of work turns downward. Thereafter, if we omit the Christmas hump and use the dotted line, the temperature and the amount of work decline together until they reach the lowest point in January. It is worth while once more to call attention to the somewhat surprising fact that in southern New England, contrary to our ordinary opinion, low temperature seems to be much more injurious than high.

This by no means indicates that high temperature is favorable. Let us consider the effect of the high temperatures of the four successive summers shown in Figure 1 (page 84). Compare

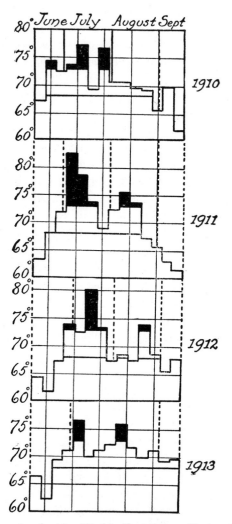

Figure 5. Average Weekly Temperature During the
Summers of 1910-13 in Connecticut

the summer dip in the Connecticut curve, that is, the area below the horizontal lines, with the heavily shaded areas of Figure 5, which shows the average temperature each week during the four summers from 1910 to 1913. The black portions indicate weeks having an average temperature night and day of over 73°. The size and distribution of these periods of extreme heat are in close correspondence with the amounts by which the curves of Figure 1 drop below the horizontal lines during the summers. This is illustrated in the following little table. The line marked "deficiency in work" indicates the amount by which the efficiency of the operatives diminished because of the hot weather, that is, the area below the horizontal lines of Figure 1. The year when the diminution was greatest is reckoned as 100 and the others in corresponding ratios. The other numbers show the area of the heavy black shading in Figure 5 and represent the intensity and duration of the hot weather. Here, too, the year of maximum heat is represented by 100, and the others by proportional values.

Year.	1910	1911	1912	1913
Deficiency in work,	58	100	8	2
Severity of heat,	52	100	50	34

In each case 1911 stands highest, 1910 next, and then 1912 and 1913. In 1911 the heat not only was extreme, but lasted long, three weeks at one time and two at another. The death rate for July, 1911, in Massachusetts was 50 per cent greater than in the preceding June. In 1910 the hot weather was not so severe, it lasted four weeks instead of five, and was divided into three parts instead of two. In 1912 the number of hot weeks was the same as in 1910. One was extremely hot, but the rest were not bad. Moreover, they did not come together, and the last was separated from the others by three cool weeks during which people had time to recover, which was not the case in 1910.

Finally, 1913 was a very mild year with only two extreme weeks which were separated by three moderate weeks.

An examination of Figure 5 makes it clear that only the extreme weeks are harmful. Thus 1911 was a truly terrible summer and 1913 a delightful one. Yet during 1911 the temperature remained above 69° for only eight weeks while in 1913 it remained above that figure for twelve weeks. Thus it appears that if the average temperature does not rise above about 70°, and if the noon temperature rarely exceeds 80°, the physical capacity of European races in the United States does not suffer any serious diminution. A slight further rise however—only four or five degrees—produces disastrous consequences. A single week of such weather does no great harm, but when several weeks come together people rapidly become weakened. The weakening is greater than appears in our diagrams, for during hot spells many of the operatives, particularly the girls, stop work entirely or stay at home in the afternoon. Those who remain are the stronger ones, and naturally their wages are higher than the general average. Moreover, in 1911 the heat was so intense that the factory shut down for two or three days. Thus, if allowance is made for these facts, the difference which a few degrees make between two summers such as 1911 and 1913 becomes even more pronounced. The full effect of a hot summer, especially when it is very damp, may be gauged by the death rate in Japan (page 95). September is there 18 per cent worse than the average, instead of 3 per cent better as in New York.

The relation between the temperature and the amount of work in winter during the four years under discussion is not so pronounced as in summer, but can easily be detected. The hot summer of 1911 was followed, as frequently happens, by an uncommonly cold winter. The reason for both is the same. Usually hot weather in New England is commonly due to the movement of heated air from the interior toward the coast, particularly from the southwest to the northeast. Cold winters are due to a similar transportation of air from the interior, this time from the north-west. The interior of a continent, as is well known, cools off very rapidly in winter and becomes hot rapidly in summer. When these conditions are carried from the interior to the coasts, they bring to New England what climatologists call a continental climate instead of the more maritime climate which otherwise prevails.

The effect of the cold winter of 1911-1912 can easily be seen in the curve for 1912 in Figure 1 (page 84). That year the average temperature where the factories are located was 19.0° for the first five weeks compared with an average of 32.7° for the three other years whose curves are given. For the next five weeks the temperature was 24.4° compared with 35.3°. The effect of this is seen in the low position of the 1912 curve of work far into the spring. The fact that the energy of the operatives remained low after the temperature began to rise suggests that the effect of extreme conditions may last long after more normal conditions begin to prevail. The same thing is suggested by the fact that after the summer of 1911 the curve of work does not rise so high in November as in the preceding May. During each of the other three years the November maximum is higher than its predecessor. Although a single winter and a single summer are not enough to prove that the effect of extreme conditions does thus persist for many months, they suggest that a long stay in an adverse climate may produce results which last for years. In spite of a previous statement, it appears that our plan of escaping from possible extreme heat by taking summer vacations in the mountains or at the seaside is wise. Equally wise is the growing habit of getting away from the severe cold for a while in winter. The only trouble is that those who most need such a change are rarely the ones who get it. If people could spend the summer on the Maine coast, the winter in Georgia, and the rest of the year in New York, they ought to be able to do the best

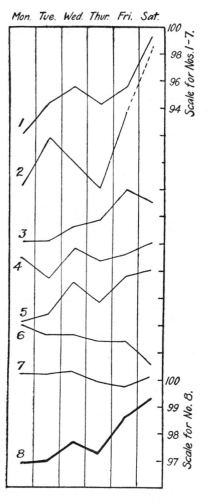

Figure 6.   Effect of the Days of the Week on Piece-Workers

1. 60 Men, April-July, 1912.
2. 60 Men, August-November, 1911.
3. 49 Girls, 1912.
4. 31 Men, 1912.
5. 14 Girls, 1912.
6. 24 Men, 1912.
7. 60 Men, January-March, 1910.
8. Weighted Average of Nos. 1-7, or approximately 230 People for One Year.

kind of work at all seasons almost without the necessity of a vacation.

The effect of temperature may be shown in more ways than have yet been presented. Let us determine how fast people work on days having various temperatures, no matter in what month they occur. The very cold days, of course, all come in winter, but may be in December, January, or February. The very hot days come anywhere from May to September, while days with a temperature of about 50° occur in almost every month of the year.

The method can be illustrated by taking all the Mondays, all the Tuesdays, the Wednesdays, and so forth, and averaging the work of each day of the week. This has been done for 230 people. The results are shown in Figure 6, which is inserted to show exactly how our results are obtained, and how necessary it is to have a large number of people. We are striving to separate the effects of one single condition from those of a vast number. We start with the wages of individuals which vary from day to day for hundreds of reasons wholly unconnected with the day of the week or the weather. The variations are so great that even if a man is influenced by the approach of pay-day, for example, we should probably not be able to detect it if we merely looked at his wages for a month or two. Therefore we average all the people of a department together, and obtain results such as appear in Figure 7. This shows the actual wages—in percentages of the maximum—which were earned by 170 people divided into five departments during five weeks in January and February, 1913. There is little uniformity in the different lines. Where one goes up the other goes down. Yet closer examination shows that in at least four out of the five departments the wages during the last two weeks were a little larger than during the earlier weeks. The variations of the different curves are in part due to the persistence of individual vagaries which have not yet been averaged out, and in part to conditions affecting whole depart-

ments. For example, a foreman is cross one day and good-
natured the next; a belt breaks and delays work; or some of the
operatives converse so much that their work suffers appreciably.
If a number of departments are averaged together these acci-

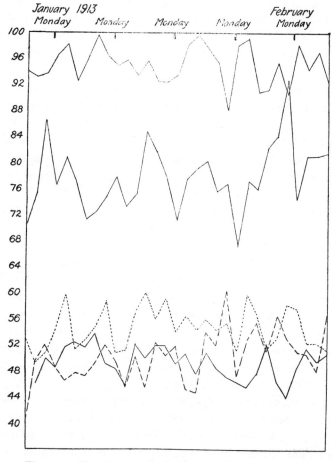

Figure 7. Variations in Daily Wages. Five Departments
(170 people) at New Britain, Conn.

dents, as well as those which pertain to individuals, disappear, but not until a great many people are considered.

To find the effect of the days of the week, we take data such as are illustrated in Figure 7, select all the Mondays, Tuesdays, and so forth, and average each day. This gives the curves of Figure 6. Here we begin to detect a certain degree of uniformity, although the accidents and peculiarities of each department are still in evidence. On the whole, however, the curves are higher at the end of the week than at the beginning. All, to be sure, are irregular, and the two lower—not counting the heavy line— slope in the opposite direction to the rest. The fact that the remaining five slope in the same direction shows, however, that these different people in different factories and during different years were subject to a common influence. Finally, we average all the departmental curves, giving each a weight proportional to the number of operatives. Thus we obtain the heavy lower line of Figure 6. This is still irregular, for although 230 people are included, all influences other than that of the days of the week are not yet eliminated. Nevertheless, the wages clearly increase toward the end of the week. If the operatives were paid by the day instead of by the piece, this would probably not be the case. They would work slowly at the end of the week by reason of being tired. With the piece-workers, on the contrary, other considerations are dominant. If they work a trifle slowly on Monday, they can make it up tomorrow. On Tuesday they can be slow and make it up on Wednesday, but a few who fell behind on Monday are beginning to work harder. So it goes from day to day until on Friday and especially Saturday many feel that their earnings for the week are insufficient, and hence make an extra effort. In some cases this may not be true, as in the curve next above the average curve. Yet it remains a general truth, and the lower curve of Figure 6 is a concrete expression of the fact that in the factory under discussion there is a difference of at least 2 per cent between Monday and Saturday. Possibly

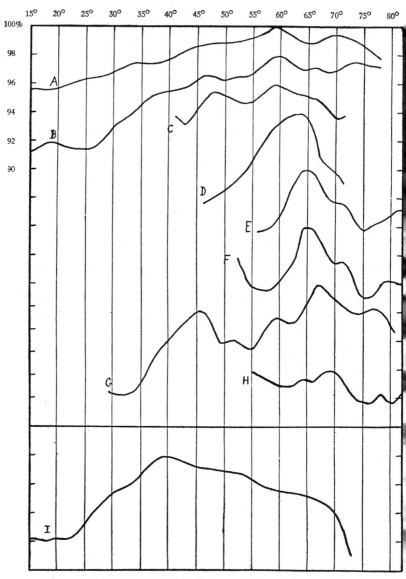

Figure 8. Human Activity and Mean Temperature

A. 300 Men in Two Connecticut Factories, 1910-13.
B. 196 Girls in One Connecticut Factory, 1911-13.
C. One Man (P) in Denmark, June-December, 1906.
D. One Man (L) in Denmark, June-December, 1906.
E. 380 Cigar-makers in Factory B at Tampa, Fla., 1913.
F. 400 Cigar-makers in Factory A at Tampa, Fla., 1913.
G. 3 Children Typewriting in New York, 1905-6.
H. 380 Cigar-makers in Factory B at Tampa, Fla., 1912.
I. 1560 Students in Mathematics and English at West Point and Annapolis, 1909-1913.

NOTE. All the curves except G and I are drawn on the same scale. The maximum in every case is reckoned as 100.

the real difference is greater, and is obscured by other circumstances. In the cigar factories of Florida it rises to a far greater value, for the Cubans are much disinclined to work after a holiday. Not only are about 10 per cent of the operatives absent on Mondays, but those who are present come so late or are so indisposed to work that they accomplish only about 80 per cent as much work as on other days. This is so important a matter that allowance for it has been made in computations where individual days rather than weeks are concerned. The figures for each day of the week for 780 men at Tampa are as follows: Monday, 81.9 per cent; Tuesday, 98.7 per cent; Wednesday, 99.8 per cent; Thursday, 100 per cent; Friday, 98.3 per cent; and Saturday, 97.9 per cent. The other days are reckoned as of equal weight, but the figures for Monday have been increased in the ratio of 82 to 100.

By the employment of a method similar to that used with the days of the week we obtain the curves shown in Figure 8. These are based on varying numbers of people, from one to over 700. Yet all show the same general character. With the exception of G and H, which are distinctly the least reliable, the physical group all reach maxima at a temperature between 59° and 65°. Even the two less reliable curves reach their maxima within the next four degrees. All the curves decline at low temperatures, that is, on the left, and also at high. The irregularities at the extreme limits are largely due to the fact that there the number of days is so small that exact results cannot be hoped for.

Figure 8, with the brief statements which accompany the respective curves, tells the whole story so plainly that it scarcely seems worth while to amplify it. Several points, however, may well be emphasized. For instance, below a certain temperature, which varies from curve to curve, a further reduction does not seem to produce much effect. People apparently become somewhat hardened, or else the conditions within the warmed houses do not change much in spite of a change in the outside air. Another noticeable thing is that the curve for girls has greater amplitude than that for men in the same region. Part of this is due to the inclusion of the group of Italians, already referred to, who are engaged in drawing hot brass and hence are benefited by the coldest kind of weather. Even if they were omitted, however, the girls' curve would still vary more than that of the men. This seems to indicate that either because of their sex or because of their age, girls are more sensitive than men.

Another point brought out by the curves is that as we go to more southerly climes the optimum temperature of the human race becomes higher. It is important to note, however, that the variation in the optimum is slight compared with the variation in the mean temperature of the places in question. For instance, in Connecticut the optimum seems to be about 60° for people of north European stock. This is about ten degrees higher than the mean temperature for the year as a whole. In Florida, on the other hand, the optimum for Cubans is about 65°, which is five degrees *lower* than the mean temperature for the year at Tampa. In other words, with a difference of twenty degrees in the mean annual temperature, and with a distinctly northern race compared with a southern, we find that the optimum differs only about 5° F. This seems to mean that for the entire human race the optimum temperature probably does not vary more than ten or fifteen degrees.

We have not yet pointed out all the important matters suggested by the curves of Figure 8. Above the optimum the curves in general begin to decline quite rapidly, but then cease to do so and at high temperatures are not so low as would be expected. This is largely because in hot weather many operatives, especially the girls and the Cubans, do not feel like work, and so stay away from the factories. Those who come in spite of the heat are the strongest and most efficient. Naturally, their average wages are higher than those of the ones who stay away, and hence the general level of our curves is too high in the portions based on the hottest weather. The mental curve, however, falls off very rapidly at high temperatures. This is because the students are obliged to be present on hot days just as on others. They must recite whether they wish or not. Hence, their curve is more reliable than the others. In this connection some experiments carried on by the New York State Commission on Ventilation are of interest. In an attempt to determine the most favorable conditions of ventilation the Commission placed several groups of persons in rooms where the temperature and humidity were under exact control, and measured their strength, mental activity, food consumption, and other conditions. The experiments lasted six or eight hours a day, and each set of subjects was tested for several weeks. Three temperatures were used, namely, 68°, 75°, and 85°. No appreciable effect upon strength could be detected, nor upon mental activity, and various other functions. This is probably because the experiments were not sufficiently prolonged. That is, the subjects were in the experimental rooms only a third or a quarter of each day, and hence their condition did not have time to change appreciably. Although the subjects did not lose in actual strength, however, their *inclination* to work declined at high temperatures even within six or eight hours.

Thus far we have been dealing with large bodies of people. It is peculiarly important to find that no matter how small the number, the same relation to temperature is discernible. One of the curves in Figure 8 shows the speed and accuracy of three children who wrote upon the typewriter a few stanzas from the "Faerie Queen" or a page from George Eliot daily for a year, and weekly for another year. Their records were kindly placed at my disposal by Prof. J. McK. Cattell. I have corrected them for the effects of practice, and have combined speed and accuracy in such a way that each has the same weight. At one period, for some unknown cause, the efficiency of the children declined greatly for two months or more. If this were eliminated their maximum would come at a lower temperature than now appears, probably not much above 60°. In the curves of individuals, we are fortunate in having careful tests made by two psychologists, Lehmann and Pedersen, at Copenhagen. They tested their own strength daily with the dynamometer, and their curves, copied directly from their monograph, are before us. One is uncommonly regular with a maximum at 64°. The other, less regular, has its maximum at 59°. The agreement of Danish curves based on single individuals with New England curves based on hundreds is highly important.

The last thing to be considered in Figure 8 is the mental curve at the bottom. It is based on so large a number of people, and is so regular, that its general reliability seems great, although I think that future studies may show the optimum to be a few degrees higher than is here indicated. It agrees with the results of Lehmann and Pedersen. Furthermore, from general observation we are most of us aware that we are mentally more active in comparatively cool weather. Perhaps "spring fever" is a mental state far more than a physical. Apparently people do the best mental work on days when the thermometer ranges from freezing to about 50°—that is, when the mean temperature is not far from 40°. Inasmuch as human progress depends upon a coördination of mental and physical activity it may be that the greatest total efficiency occurs halfway between the mental and physical optima, that is, with a mean temperature of about 50°.

Curves such as those of Figure 8 are not peculiar to man alone. They are apparently characteristic of all types of living creatures. To begin with plants, many experiments have determined the rate of growth of seedlings at various temperatures. The commonest method has been to grow different sets of seedlings in large numbers under conditions which are identical except in temperature, and then to measure the average length of the shoots. In all cases growth is slow at low temperatures, increases gradually with higher temperatures, reaches a maximum like that of man, and then falls off quickly. The course of events, however, is not always so regular as here indicated. The curve of wheat, for example, as worked out by MacDougal is given in Figure 9. The peculiar double maximum there seen appears in each case where careful tests are made. It seems to be due to some inherent quality of the plant, and is of especial interest in

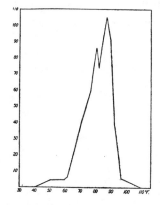

Figure 9.   Growth of Wheat at Various Temperatures
After MacDougal

The figures on the left indicate growth in mm. during 48 hours

our present study because we shall soon come upon an analogous case in man. When many species are averaged, such irregularities disappear, and we obtain the curve at the bottom of Figure 10, which has been prepared by MacDougal on the basis of his own measurements and others given in such works as Pfeffer's *Physiology of Plants.* Many of the lower plants, such as marine algæ, have their optima at lower temperatures than those here indicated, and the same is probably true of Arctic species. On the other hand, certain low algæ which grow in hot springs must have their optimum at a temperature above that of ordinary plants. These differences are immaterial. We are now concerned only with the fact that so far as plants have been measured, their response to temperature resembles that of man.

Apparently, we have to do with a quality which pertains to all kinds of living beings, and is presumably an inherent characteristic of protoplasm. The nearest approach to pure protoplasm is found in unicellular organisms whose bodies show only the beginnings of differentiation into parts having separate functions. The infusoria furnish a good example. One of these, paramœcium, has been carefully studied by L. L. Woodruff. His

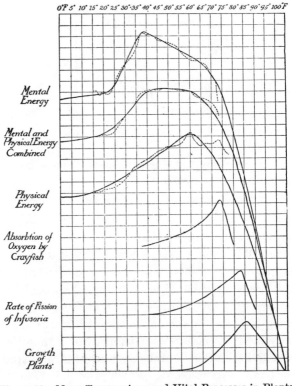

Figure 10. Mean Temperature and Vital Processes in Plants,
Animals and Man

original purpose was to determine whether it was possible for this organism to keep on reproducing itself without conjugation for any great length of time. Under the conditions of nature the small motile cells often spontaneously develop a median cell wall and ultimately divide into two new individuals, thus reproducing the species. This process, however, does not go on indefinitely, for when two cells come in contact they fuse with one another, and then begin a process of fission which, like the other process, ends in two individuals. Thus we have two types of reproduction, asexual and sexual, which apparently give rise to the same kind of paramœcia. Woodruff's purpose was to determine whether asexual reproduction can persist indefinitely, or whether it leads in time to extinction. He has shown that if the media of nutrition contain a sufficient number of elements, paramœcium can reproduce itself indefinitely by the asexual method. Between May 1, 1907, and May 14, 1914, he had carried his cultures through 4417 generations without conjugation. In the spring of 1924 the paramœcia were still thriving after about fourteen thousand asexual generations. In the course of this work he has found that the rate of cell-division is an accurate test of the conditions under which protoplasm exists. For example, when extracts from nephritic kidneys or certain other diseased organs are added to the nutrient solution, even though they are present in such small quantities that they cannot be detected by chemical analysis, they make their presence evident by a falling off in the rate of fission.

One of Woodruff's most important lines of work has been to test the relation of his infusoria to temperature. From many experiments he finds that their activity corresponds closely to van't Hoff's law of chemical activity. According to this well-established law, chemical reactions of most kinds at ordinary temperatures become nearly three times as active with every rise of 10° C. Even in inorganic chemical reactions, however, and far more in those of the living cell, there is a distinct limit where this rule breaks down. This limit forms the optimum of the species. At higher temperatures the degree of activity declines, and finally death ensues. On the basis of these conclusions, Woodruff's data permit us to draw the second curve from the bottom in Figure 10.

The next higher curve shows the amount of oxygen absorbed by the common crayfish at various temperatures. The most extensive work on this subject appears to have been done by Brunnow. The facts here given are taken from the summary by Pütter in his *Vergleichende Physiologie*. The amount of oxygen absorbed by an animal is an excellent measure of its physical activity. When supplemented by measurements of the amount of carbon dioxide given off, and of the speed with which certain other metabolic or katabolic processes take place, it gives a true picture of the animal's general condition. Apparently, these various processes follow van't Hoff's law just as do the growth of plants and the cell-division of the infusoria. The optimum in the three cases does not vary greatly, that for plants being about 86°, for paramœcium 83°, and for the crayfish 74° F.

Physiologists are not yet fully agreed as to the cause of the phenomena shown in these curves, although there is little doubt as to the general facts that they imply. One hypothesis may be briefly stated. According to Pütter's summary, the most probable explanation is that activity goes on increasing according to the ordinary chemical law until it becomes so great that the organism is not capable of absorbing the necessary oxygen. That is, at a low temperature the creature easily gets what oxygen it needs, and gives it out again in the form of carbon dioxide or of other oxidized products which remove the waste substances from the body. As the temperature rises, the normal increase in chemical activity takes place, the animal is still able to get rid of all its waste products, and thus its life processes are strengthened. With a further rise of temperature a change sets in. The chemical processes which break down the tissues of the body become still more active, but the supply of waste products to be eliminated by oxidation becomes so great that they cannot all be removed. This is because in every organism there is a distinct limit to the amount of oxygen which the creature can mechanically convey to different portions within a specified time. If the supply of oxygen is not sufficient to oxidize all the waste products, some of these will remain in the system. They act as poisons. Their first effect is to diminish the organism's activity. If they accumulate to too great an extent death ensues.

The discussion of this hypothesis must be left to the physiologists. They must decide whether the hypothesis which explains the curves of cold-blooded animals and plants is also applicable to warm-blooded animals. There can be little doubt, however, that variations in the rate at which metabolism takes place in the human body play a part in the variations in efficiency which we are here studying. The researches of Thomson illustrate the way in which we are beginning to discover the truth. In Manchester, England, from April to July, 1910, and again in March, 1913, he measured the percentage of CO_2 given off in the breath of four individuals under different conditions of temperature, humidity, and pressure. From his figures, given in the *Manchester Memoirs*, I have compiled the following tables:

I. Percentage of CO_2 Exhaled by Four Persons Under Different Conditions of Temperature

Temperature,	50-51°	52-53°	54-55°	56-57°	58-59°	60-61°	62-63°	64-65°	98° Blood Heat
Percentage of CO_2,	4.77	4.72	4.66	4.71	4.61	4.41	4.38	4.62	4.80+?

II. Percentage of CO_2 Exhaled by Four Persons Under Different Conditions of Humidity

Relative Humidity,	70-75%	76-80%	81-85%	86-90%
Percentage of CO_2,	4.75	4.60	4.60	4.45

The interpretation of these tables is difficult, and I can merely offer a suggestion. An increase in the proportion of CO_2 exhaled from the lungs obviously indicates an acceleration of the metabolic processes which break down and consume the bodily tissues. This liberates energy which may manifest itself in at least three ways and possibly more. It may give rise to heat which is used to maintain the body at the normal temperature; it may be used to accomplish physical or mental work; and it may cause an excess of heat which gives rise to further metabolism of a harmful nature. In the first part of the table the percentage of CO_2 is comparatively high at the lowest temperature recorded by Thomson, and decreases with only slight irregularity till the thermometer reaches 62° F. This is close to the temperature which we have found to be the optimum. Below that point the increased metabolism is probably needed to keep the body warm. At higher temperatures increased production of CO_2 is again apparent. This perhaps means that too much chemical activity is taking place, and that toxic substances are accumulating in the way suggested by Pütter. At the optimum, according to this interpretation, the body does not have to use an undue portion of its strength in keeping warm, nor is it injured by too great stimulation. Thus it is in the best condition for work.

The second part of the table shows that in the driest weather which England enjoys, metabolism is more active than in wet weather. Perhaps part of this is due to the fact that in dry air the body loses water and is cooled by evaporation, and hence requires more heat than in wet air of similar temperature. There is more to the matter than this, however, but further measurements are needed before an adequate explanation can be offered. All that can be done here is to point out the fact that in man, as in the lower organisms, activity varies according to temperature. This is evident in Figure 10, where the dotted upper line is the curve of mental activity, while the accompanying solid line shows the conditions if all accidental irregularities could be removed. The third line in the same way represents the physical activity of both men and women in Connecticut. I have not used the figures from the South because they are not quite so reliable as those from Connecticut. Finally, the second line from the top shows physical and mental activity combined, each being given the same weight. It may be taken as representing man's actual productive activity in the things that make for a high civilization. The resemblance of the human curves to those of the lower organisms is obvious. In general, the lower types of life, or the lower forms of activity, seem to reach their optima at higher temperatures than do the more advanced types and the more lofty functions such as mentality. The whole trend of biological thought is toward the conclusion that the same laws apply to all forms of life. They differ in application, but not in principle. The law of optimum temperature apparently controls the phenomena of life from the lowest activities of protoplasm to the highest activities of the human intellect.

CHAPTER VI

WORK AND WEATHER

THE effect of a given climate depends on two primary factors. One is the character of the seasons as expressed in averages such as are furnished by our weather bureaus. The other is the changes from day to day, that is, the weather. The boy quoted by Mark Twain was nearly right when he defined the difference between weather and climate as being that "Climate lasts all the time and weather only a few days." Two climates may be almost identical in their seasonal averages, and yet differ enormously in their effect on life, because in one the change from day to day is scarcely noticeable, while in the other there are all sorts of rapid variations. The old Irishwoman who was driving her pigs to market in a pouring rain did not realize it, but she gave expression to a truth of the greatest importance, when a friend pitied her for being out in such weather, and she replied, "Indade it's bad, but sure it's thankful I am to have any kind of weather."

The changes from one day to another depend largely upon our ordinary cyclonic storms. In such storms the barometer goes down and then up; the wind changes in direction and velocity; the air becomes humid, clouds gather, rain usually falls, and then clear skies and dry air prevail; the temperature also changes, often rising before a storm and falling afterward, although the exact sequence depends on the location of a region in respect to the ocean and to the center of the storm; the daily range of temperature also varies, for in damp or cloudy weather the nights do not become so cool nor the days so warm as when the air is clear. To understand the influence of the weather all these conditions must be investigated. Most of them, however, appear to be of relatively slight importance when considered by themselves. For instance, Lehmann and Pedersen could find no appreciable effect of the pressure of the atmosphere except where low pressure prevails a long time. The decrease in efficiency at such times, however, is probably due more to prolonged cloudiness and its attendant circumstances than to the barometric conditions. My own work leads to the same result. The curves of efficiency compared with pressure are so contradictory that it does not seem worth while to publish them. The same is true of the range of temperature from day to night, and of the direction and force of the winds. I have no doubt that all these matters are important, and that some day their effect will be worked out. In general, however, their influence is exerted indirectly through changes in temperature and humidity. In hot weather a great range from day to night is unquestionably highly favorable, but at ordinary temperatures it seems to make no special difference, except through its effect upon the mean temperature.

As to the winds, Dexter, in his book on *Weather Influences*, shows that they produce a marked effect upon the nerves, as is indicated by the unruliness of school children in Denver when high south winds prevail. Part of this is doubtless due directly to the wind, but the unseasonably high temperature and extreme dryness which accompany it are probably more important. Yet we are all conscious of the effect of a steady high wind. Some people are stimulated. I have seen a small boy, who was usually very quiet, climb to the top of a tall tree when a violent wind came up, and swing in the branches, singing at the top of his voice. For a while such stimulation is probably beneficial, but if continued day after day it makes people excitable and cross. A striking example of the effect of a prolonged wind ·is seen in eastern Persia in the basin of Seistan. During the summer, from

June to September, the so-called "Wind of One Hundred and Twenty Days" blows so violently from the north that in the oases trees cannot grow except under the lee of high walls. The acrid wild melon, which ripens its beautiful little green and yellow fruit in the desert, does not spread its slender branches in all directions after the common fashion of plants. The gales crowd the branches into a sheaf which points so uniformly in one direction, a little to the west of north, that it can safely be used as a compass. When Europeans have to endure this wind they say that it is one of the most trying experiences imaginable. Not only does it render them irritable, but it deadens their initiative and makes them want to stay idly in the shelter of the house. The natives, although possessed of many good qualities, are inert and inefficient even in comparison with their fellow Persians who live farther to the north and west. On the whole, we may probably conclude that occasional short-lived gales and frequent light or moderate winds are beneficial, while long periods either of steady calms or of gales are depressing.

Aside from the conditions of weather already mentioned, there are two whose effect appears plainly when curves are constructed according to the method described above. One is the change of temperature from one day to another, and the other is the character of the day as to clouds and sunshine. In considering changes of temperature from one day to the next, we deal with the mean temperature for each day and not with the extremes. A change of as much as 15° is rare. Suppose that the thermometer stands at 60° at sunrise, rises to 80° by two o'clock in the afternoon, then falls rapidly to 50° at sunset and to 40° by midnight. Suppose also that the next day the temperature is 40° at sunrise, rises a little above 55° during the day, and falls again to 45° at night. The two days would be very different, and we should speak of them as being marked by a very great change of temperature, a difference of 40° within ten hours. Yet the average of the first day would be about 64° and of the second 49°, a difference of only 15° in the mean temperature.

On the basis of this supposition the reader can estimate the importance of the various degrees of change indicated in Figure 11. At the left the curves show the average efficiency on days when the temperature has fallen; in the middle are the days with no change; and at the right are the days characterized by a rise. Taking only the two upper curves, those for men and girls in Connecticut factories, the resemblance is striking. When we consider the heterogeneous character of the original materials the resemblance is still more important. The men's curve is based on 120 men at Bridgeport in 1910 and 1911, and on 180 men at New Britain in 1911, 1912, and 1913. The girls' curve is based on 196 girls at New Britain in 1911, 1912, and 1913, and on 60 girls at New Haven in 1913 and 1914. Even when the girls and men are working in the same factory, there is no reason, aside from the weather, why their wages should be high on the same day. The chief difference between the two curves is that the one for the girls varies more than that for the men, and reaches its maximum slightly farther to the right. Apparently, here, just as in the case of mean temperature, the girls because of their age or sex are more subject to the influence of the weather than are the men, and hence their curve dips deeper.

Let us now interpret the upper curves, beginning at the middle. There they fall to their lowest level. This means that when the temperature of today is the same as that of yesterday, people work more slowly than after a change, no matter whether the change is upward or downward. A variable climate is therefore highly desirable if people are to be efficient. Perhaps the most surprising feature is that the lowest point of the physical curve, and a depression of the mental curve, C, come not at 0°, but at −1°. The zero point is low, lower than any point of the physical curves except −1°. Hence, our conclusion as to the injurious effect of uniform temperature is justified, but that

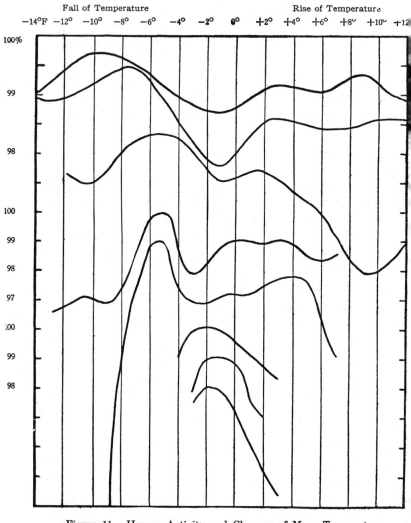

Figure 11   Human Activity and Changes of Mean Temperature
from Day to Day

A. 300 Men in Two Connecticut Factories, 1910-13.
B. 256 Girls in Two Connecticut Factories, 1911-13.
C. 460 Students in Mathematics and English at West Point and Annapolis, 1909-1913.
D. 760 Cigar-makers at Tampa, Fla., in Winter (October-March), 1912 and 1913. Factory A.
E. 400 Cigar-makers at Tampa in Winter, 1913. Factory B.
F. 400 Cigar-makers at Tampa in Summer (April-September), 1913. Factory B.
G. 380 Cigar-makers at Tampa in Summer, 1912. Factory A.
H. 380 Cigar-makers at Tampa in Summer. 1913. Factory A.

does not explain the curious dip at $-1°$. The repetition of the same phenomenon in each of the three upper curves, and a similar occurrence at $-2°$ and $-3°$, respectively, in the two curves for the winter in Florida strongly suggest that we are confronted by a peculiarity which pertains to man as a species, in the same way that a double optimum of mean temperature pertains to wheat as shown in Figure 9. Possibly, a slight fall in temperature causes people to shiver, as it were, and only when the fall is slightly larger is the circulation of the blood so stimulated as to increase the activity of the various organs. In the South it may be that people's blood is more sluggish than in the North, so that the reaction due to cooler weather does not follow quite so soon, and hence the period of shivering is not over until the fall in mean temperature amounts to more than about $3°$. I do not assert that this is so, but it is the only explanation that comes to mind.

To go on with our interpretation of the physical curves, a slight rise of temperature seems to be favorable, but beyond that the favorable effects of increased heat, which are strong in cold weather, are neutralized by the unfavorable effects in warm weather. In fact, our personal experience tells us that even when the heat is not extreme, a sudden rise may make us uncomfortable and lazy, as often occurs in the spring. In spite of this, however, a rise is in general better than uniformity. When the temperature falls, on the other hand, a distinct stimulus is received, provided the fall amounts to as much as $4°$. The best effects are seen with a fall of from $6°$ to $9°$ with girls and of $7°$ to $11°$ with men. Here again the implication is that men are on the whole less sensitive than girls. An extreme drop is not so favorable as one of more moderate dimensions, especially for the girls. Taking the physical curves as a whole, the greatest amount of energy would be expected in climates where the mean temperature first rises $2°$ or $3°$ a day for a few days and then drops $4°$ to $8°$ a day. If the changes are greater than this, the effect is still stimulating, but not so beneficial as under the more moderate conditions. If there is practically no change, on the contrary, the level of efficiency lies within the low central depressions of our curves, and is less than under either of the other conditions.

Mental work resembles physical, but with interesting differences. When the temperature falls greatly, mental work seems to suffer more than physical, and declines as much as when there is no change. It receives a little stimulus from a slight warming of the air, but appears to be adversely affected when the air becomes warm rapidly. This last statement, however, must be qualified. The physical curves are based on the complete year, and the conditions of summer have an opportunity to balance those of winter. The results show the net effect for all seasons combined. The mental curves, on the other hand, do not include the summer vacation, which lasts from the middle of June to the first of September at West Point, and from the middle of May to the first of October at Annapolis. If this were included, the effect of a pronounced lowering of the temperature would be more noticeable than at present, for such a lowering is naturally more stimulating in July than in January. In another respect, also, the curve of mental efficiency needs modification. It is based on figures from two climatic provinces, namely, southern New York and Maryland. The great decline at times when the temperature rises rapidly is due largely to conditions in Maryland, where the hot days of the spring are much more debilitating than in New York. The students belong to a race which has never learned to endure sudden heat. Hence they feel it strongly. If allowance is made for the two conditions just mentioned, the mental curve will approach much more closely to the physical. A drop of temperature amounting to 8° or more will appear more stimulating than now seems to be the case, and a rapid rise will not seem so harmful. Hence, the general conclusion for both physical and mental activity will be essentially the same.

It may be summed up thus: Taking the year as a whole, uniformity of temperature causes low energy; a slight rise is beneficial, but a further rise is of no particular value; the beginning of a fall of temperature is harmful, but when the fall becomes a little larger it is much more stimulating than a rise; when it becomes extreme, however, its beneficial qualities begin to decline. This conclusion must, of course, be appropriately modified according to the season. A cold wave in January is very different from one in July. In our curves we have given January and July an opportunity to neutralize one another. They have not done so. This means that after all allowances have been made for the seasons, the total effect of cold waves is decidedly beneficial, and of warm waves slightly so. Frequent changes, therefore, are highly desirable.

Let us pass on now to the Florida curves. Here we find a curious difference between summer and winter which is not easy to understand. Let us leave that for the moment, however, and consider only the two winter curves. Their general resemblance is marked. The differences at the extremities are not important because the number of days there concerned is very small. It must be remembered that the two curves are from independent and rival factories. The position of any particular point in either curve depends upon a number of days scattered irregularly through the months from October to March. Aside from a genuine effect of climate, there seems to be no possible way in which 400 men in one factory in 1913 could be made to work so that their curve would be the same as that of 380 men in another factory in the two years 1912 and 1913. Here, as in Connecticut, West Point, and Annapolis, we are apparently dealing with a peculiar quality which is inherent in the human species.

One of the Florida curves, E, is low at 0°, while the other is medium. This means that days when there is no change of temperature are not particularly favorable. At plus 2° to plus 4°, however, both are fairly high, which indicates that a moderate rise of temperature is favorable. A further rise seems to be harmful. The effect of a slight fall of the thermometer has already been discussed. A further fall is beneficial. The most notable thing about curves D and E is the maximum from $-4°$ to $-7°$. It comes at about the same place as the mental maximum, and is similar to the Connecticut maximum except that the people in the far South do not seem to be able to stand such extreme changes as do those in the North. In fact, it seems most significant that the Connecticut men, who are the strongest of our various groups, are most stimulated by a strong change of temperature. The Connecticut girls come next, but, being less sturdy, they do not profit quite so much by rigorous conditions. The mental curve is largely determined by Annapolis, and as the climate there is less severe than in Connecticut, the students seem to feel more keenly the effects of extreme changes, although they are stimulated by those of moderate dimensions. The same is still more true of the people of Florida in winter. Finally, during the summer the Floridans are stimulated by a slight drop of temperature, provided it is not enough to make them feel chilly, but enough to start their blood in motion. A greater drop makes them feel cold, while even the slightest rise of temperature in their long monotonous summer is unfavorable.

We are ready now to sum up our results. The outstanding point is that changes of temperature, provided they are not too great, are more stimulating than uniformity, while a fall is more stimulating than a rise in the latitudes now under consideration. The effect of changes depends largely upon the degree to which people are inured to them. When they are weakened by a long hot period like that of the Florida summer, even a slight cooling of the air brings relief and activity, provided it does not go so far as to make people feel chilly. When the same Floridans become wonted to the somewhat sterner, albeit mild air of their winter, the first effect of a lowering of the temperature may be to make them shiver, but soon they are stimulated, and work fast. They are not so tough, however, as to be able to get benefit from the occasional days when really strong cold waves sweep down upon them. On the other hand, a rise of temperature stimulates them, unless it is of considerable severity. Farther north the same applies except that, being tougher, the people are more benefited by strong changes. Judging by the difference between summer and winter in Florida, it looks as if a little hardening would cause even the Cubans to respond favorably to changes at least as severe as those in Maryland, thus making the left-hand part of their curve like C in Figure 11. Taking it all in all, the one thing that stands out preëminently is that a fall of from 4° to 7° is everywhere stimulating, provided people are accustomed to it.

Man is not the only organism that is benefited by changes of temperature. Numerous experiments have shown that plants are subject to a similar influence. If a plant is subjected to unduly low or high temperature, its growth is retarded. As the temperature approaches the optimum, the rate of growth increases. When the optimum is maintained steadily, however, not only does the increase cease, but retrogression sets in, and the rate of growth declines. A moderate change of temperature away from the optimum and then back again after a few hours checks this decline, and keeps the plant at a maximum degree of activity. Thus conditions where the thermometer swings back and forth on either side of the optimum are distinctly better than where the optimum is maintained steadily. Thus it seems to be a law of organic life that variable temperature is better than uniformity.

The physiological process by which frequent changes of temperature affect the body is not yet known. The best suggestion seems to be that of Dr. W. B. James. It is universally recognized that one of the most important of the bodily functions is the circulation of the blood. The more active and unrestricted it is, the more thoroughly is the whole system nourished and purified.

Provided it does not impose an undue strain on the heart o arteries, anything that stimulates the circulation appears to be helpful. Changes of temperature are a powerful agent to this end. Witness the effect of a bath, either cold or very hot. Few things are more stimulating than a Swedish bath. An attendan holds two hoses, one with cold water and the other with hot, and plays them alternately upon the patient. A man goes into such a bath with hanging head and dragging feet. He comes out with head erect and a new spring in his walk. Apparently, frequen changes of the temperature of the air produce much the same effect. No one change produces so pronounced an effect as a Swedish bath, but the succession of stimuli due to repeated changes throughout the year must be of great importance.

Before leaving this subject, let us test the effect of change in still another way. Let us see what happens during an average series of days such as make up our common succession of weather in New England. The ordinary course of events is first a day o two of clear weather, then a day or two of partly cloudy weather next a cloudy day with or without rain, and finally another cloudy day during which rain falls. Then the sky clears in prepa ration for another similar series. On this basis I have formed the six groups indicated at the top of Figure 12. At the left, the efficiency on all clear days which follow cloudy or partly cloudy days has been plotted, just as in another diagram we plotted the efficiency on Mondays. Next come the clear days which follow another clear day. If several of these follow in unbroken suc cession, they are all included, but a third or fourth clear day is rare. In the next group come the partly cloudy days which follow either a clear or a cloudy day. The great majority follow clear days. A second partly cloudy day is much rarer than a second clear day, and a third is still rarer. The first cloudy day the fifth column, includes cloudy days which follow either clear or partly cloudy days. Finally, the sixth column includes not only the second cloudy day, but the third and fourth if such are recorded. In general, this column represents days when a storm comes to an end, while the one to the left of it represents the time when a storm first becomes well established. The rest of the diagram, to the right of the sixth column, is merely a repetition of the part already described. It is inserted to show how an ideal series of storms would repeat itself.

Figure 12 discloses some surprising facts. For instance, the first clear day is characterized by the slowest work in the two

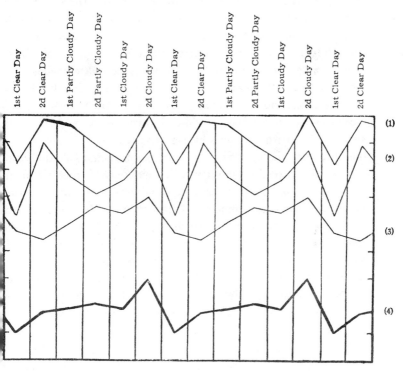

Figure 12. The Stimulus of Storms

(1) 60 Men at Bridgeport, 1910.
(2) 60 Men at Bridgeport, August, 1911-July, 1912.
(3) 170 Men and Girls at New Britain, 1913.
(4) Weighted Average of (1), (2), and (3), Equal to 290 People for One Year.

upper curves and by almost the slowest in the third. Our impression of the stimulus of the bright, clear air after a storm receives a flat contradiction. It is apparently psychological not physical. The second clear day makes a better showing than the first. It stands high in two curves, and low in only one. The first partly cloudy day is high in one curve, and medium in two. The second partly cloudy day is medium in all three. The same is true of the first cloudy day. The last cloudy day is as surprising as the first clear day. In each of the three curves it stands highest. People work fastest at the end of a storm. In the lower curve of Figure 12, the whole matter is summed up in a single line. Here we see that during an average "spell of weather" people are least efficient on the clear days; moderately efficient on the partly cloudy days, and on the first cloudy day; and most efficient at the end of a storm. We may tell ourselves that this is unreasonable, but when we think it over, we are likely to be aware of its truth. Before a storm we may feel depressed, but at the end, when the rain or snow is almost over and the air begins to have that excellent quality which makes us forget all about it, we bend to our work with a steadiness and concentration which are much less common at other times. Hellpach emphasizes this in his book on the psychological effect of geographical conditions. We fail to appreciate it largely because the æsthetic impressions of a beautiful, clear day are felt much more consciously than are the physiological conditions which throw us vigorously into our work. Each storm, with its changing skies, varying humidity, and slow rise and rapid fall of temperature, is a stimulant. Each raises our efficiency.

This ends our survey of the effect of climate upon daily work in the eastern United States. We have considered the influence of the seasons, of mean temperature, of humidity, of winds, of changes of temperature from day to day, and of the character of each day and its relation to storms. We have also seen that although different races, or people under decidedly diverse climatic environments, are at their best at slightly different temperatures, the differences are inconsiderable, and changes of temperature are as valuable to one as to the other. The question now arises whether the climatic effects are really of great importance. In Figure 12, the stimulus of the succession of clear and cloudy days amounts to only 1 per cent. In Figure 11 changes of temperature from day to day produce a variation of only a little over 2 per cent, if we omit the irregular and unreliable extremities of the curves. In Figure 4, the maximum effect of humidity appears to be only 3 per cent. In Figure 8, however, the differences are greater, for the effect of mean temperature upon the girls in Connecticut is 7 per cent. Finally, in Figure 1, the effect of the seasons reaches nearly 9 per cent when four years are averaged, and nearly 15 per cent for individual years.

These figures are far from representing the full importance of the various factors. This will readily appear from a little consideration. In the preceding paragraph, the percentages increase in proportion to two conditions, first, the degree to which the influence of a single factor is separated from the influence of all other factors, and second, the length of time during which each factor is able to exert its influence. The smallest figure, 1 per cent in Figure 12, does not represent any individual factor, unless it be cloudiness. It does not even represent the fluctuations which attend an individual storm, for the days were selected without regard to their position in a cyclonic disturbance, but simply according to their cloudiness. The variations shown in the curve are due to many factors, including mean temperature, changes of temperature, relative humidity, and others of minor importance. As no two of these are necessarily at their maximum at the same time, they neutralize one another. Moreover, a given condition lasts only a day in most cases, and so has no opportunity to produce any great effect. In the curve of changes of temperature from day to day, which shows the next larger effect, a single factor is singled out. Its full force can by no means be seen, however, for the humidity often varies in such a way as to neutralize it. Moreover, the effects of especially low or high temperatures may often completely overshadow any stimulus arising from the mere fact of a change. Furthermore, the effect of changes of temperature rarely continues more than two days. For example, if the thermometer averages six degrees lower on one day than on the preceding, it may happen that there will be a further drop before the next day, but there is far more chance that the temperature will rise a little or remain stationary, or fall so little that it will not be stimulating. Hence, the effect is rarely cumulative, and the influence of each single day must usually stand by itself. Much the same is true of relative humidity, except that by heating our houses we artificially induce long periods of great aridity. The effects of mean temperature, on the other hand, have greater opportunity to show their full importance, though they, too, are hampered. Relatively low or high temperatures last many weeks, which makes it possible for the effect of day after day to accumulate. Yet our curves by no means show the full effect, for a cold day with a mean temperature of 30° may come in November at a time when efficiency is still at its highest. It produces its normal effect, but a single unpropitious day, or even a week, does not suffice to depress people's vitality to a degree at all approaching the low limit reached after two months of cold weather. Likewise, a day with the most favorable temperature, not far from 60°, may be sandwiched between very hot days in July, or between two cold days in March. Hence, people will display little energy on those particular days, and the average efficiency at the optimum temperature will appear correspondingly lower than it ought. Finally, the seasons have more opportunity than the individual climatic elements to produce their full effect. Even here, however, the variability of our climate does not allow any special combination of circumstances to work long unimpeded. Warm waves break the cold periods of winter, and cool waves come in summer. Storms are more active in winter than in summer, and hence their stimulus works toward overcoming the effect of prolonged cold. Moreover, no single season is of great duration, and extreme conditions do not last long enough to produce their full effect. From all this we may conclude that the total influence of climate upon energy is much greater than appears in any one of our curves.

The difficulty of determining the exact proportions of any individual influence may be made clear by an example. We know that man's power to work depends upon food, drink, sleep, and clothing. Suppose that while he was still supplied with these in normal quantities we were to try to measure the effect of each. We should test his strength at stated intervals after he had eaten his meals, or after he had had a drink. We should find out how many hours he slept each night and compare that with his work. We should measure his achievements before and after he put on his spring underwear or fall overcoat. We might get results, but it is highly doubtful whether they would be as distinct as those here discussed. We have no difficulty in measuring the effect of food, drink, sleep, and clothing, for we can easily vary them to suit the needs of our experiment. With climate the case is different. We must take it as we find it, and must experiment on people who are constantly subject to its influence. Some day we shall test people first in one climate and then in another, but that will be difficult because it takes a considerable time for climate to produce its full effect. Being obliged to search for the effects of climate without being able to change them in accordance with the needs of our experiment, we are in almost as difficult a case as the experimenter who should desire to determine the effect of the amount and kind of food consumed by a group of individuals, but who had no control over how much they ate. They might allow him to measure what was set before them at each meal and what remained when it was over, but they would eat as much as they liked and when they liked. He would get results, if he did his work carefully, but they would by no means represent the full effect of food.

The influence of climate upon men may be likened to that of a driver upon his horse. Some drivers let their horses go as they please. Now and then a horse may run away, but the average pace is slow. Such drivers are like an unstimulating climate. Others whip their horses and urge them to the limit all the time. They make rapid progress for a while, but in the end they exhaust their animals. They resemble climates which are always stimulating. In such climates nervous exhaustion is likely to prevail and insanity becomes common. A third type of drivers first whip their horse to a great speed for a mile or two, and then let them walk slowly for another mile or two. They often think that they are accomplishing great things, and they are better off than the two types already mentioned, but they still have much to learn. They are like a climate which has a strong contrast of seasons, one being favorable and the other unfavorable. Still a fourth kind of driver may whip his horse sometimes and sometimes let him walk, but what he does chiefly is to urge the animal gently with the voice, then check him a little with the rein. By alternate urging and checking he conserves the animal's strength, and in the long run can cover more distance and do it more rapidly than any of the others. Such a driver resembles a climate which has enough contrast of seasons to be stimulating but not to create nervous tension, and which also possesses frequent storms whose function is to furnish the slight urging and checking which are so valuable in the total effect, although each individual impulse is almost unnoticeable.

CHAPTER VII

HEALTH AND THE ATMOSPHERE*

WE have investigated the relation of the weather to human energy. Let us do the same for health. In a previous chapter we saw that variations in the death rate in New York and Japan and the gain in weight among tubercular patients exhibit seasonal fluctuations much like those of workers in factories. As a means of measuring health, deaths are more important than disease for two reasons. First, experience has shown that when several years are averaged together, the death rate is an almost perfect measure of the number and severity of the diseases which afflict a community. Second, the records of disease are very scanty and imperfect; they have never been tabulated on any large scale for the entire population. Only in rare cases can the records of certain diseases be used as well as those of deaths. Accurate mortality records, on the contrary, have now been kept for many years.

The health of practically every community varies in response to the seasons. In the northern United States most physicians are far busier in February and March than in May and June. In July and August the demand for their services increases again, especially among children. Then come the best months of the year, especially October, when good health and good spirits abound. Different types of disease, to be sure, display different seasonal adaptations. Those of the respiratory organs, for ex-

---

* This chapter and the two that follow are entirely new. They are based largely on publications mentioned in the list at the end of the prefaces, but none of the material in Chapter IX has hitherto been published.

ample, reach a maximum in winter, while those of the digestive
tract are more numerous in summer.

Admitting, then, that both energy and health show marked
seasonal variations, our aim is to determine how closely the two
sets of variations are in harmony. Part of the answer is illustrated in Figure 13. The upper line, A, represents variations in

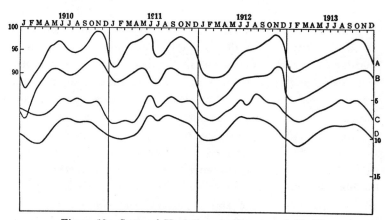

Figure 13. Seasonal Variations in Energy and Health

A. Work of factory operatives in Connecticut.
B. Work of factory operatives in Pennsylvania.
C. Health (death rate inverted) in Connecticut.
D. Health (death rate inverted) in Pennsylvania.

Scale for A and B on left, for C and D on right.   B and D are placed below A and C for convenience, although belonging at essentially the same level.

the efficiency of factory workers in Connecticut from January,
1910, to December, 1914, as already given in Figure 1. The
second line, B, is the similar curve for Pittsburgh. The two lower
lines illustrate fluctuations in health in Connecticut (C) and
Pennsylvania (D). They are the curves of the death rate inverted so that high parts indicate good conditions or few deaths,
and low parts, poor conditions or many deaths, thus per-
mitting easy comparison with the efficiency curves. It is ob-

vious not only that the two efficiency curves and the two health curves run almost parallel, but that there is also a close parallelism between health and efficiency. Aside from the weather all possible causes of such parallelism seem to be excluded, for epidemics, business disturbances, and the like, did not occur in such a way as to explain the similar fluctuations in two diverse phenomena hundreds of miles apart.

Note how closely the four curves agree even in details. Low efficiency during January, 1910, is followed in a month or two by very poor health. During the spring both efficiency and health improve. Then comes the summer with a mild tendency toward a drop in all the curves, and the autumn with the main maximum of the year. In 1911 the parallelism of the four curves is again evident, as is the lag of the mortality curve after that of efficiency. In 1912 and 1913 the sag of all the curves in summer diminishes and disappears; for those years, it will be remembered, had only short periods of really hot weather. The similarity of the four curves, especially in the summer, would be still more marked were it not that the deaths of children under two years of age have been omitted in the mortality curves.

Since many other observations point in the same direction, we conclude that unfavorable weather, such as commonly prevails in January, has an immediate effect in reducing people's vitality and energy. Hence their work falls off. At the same time they become more susceptible to disease. Accordingly, in due time the number of deaths increases. Naturally the greatest mortality lags several weeks after the lowest efficiency; it takes time for bacteria to produce infection, and for infection to lead to death. The lag is longer in winter when respiratory diseases are the chief enemy than in summer when digestive diseases with their more rapid course are the chief foes. The agreement between health and energy is thus so close that both appear to depend upon essentially the same fluctuations of the weather.

It is especially important to determine the relation of the

weather to mental as well as physical health. Hence peculiar interest attaches to certain studies of mental abnormalities carried on by Norbury.* He finds that the admissions to psychiatric hospitals show that "Mental disorders in their incipiency and recurrence parallel the efficiency curves of Huntington. (Maximum in the spring, minimum in the autumn.)" His curves show that certain maxima of admissions for mental disorders occurred at the following periods:

Civil Hospitals of New York State, 1916-1921, June.
Norbury Sanatorium in Jacksonville, Illinois, 1900-1923, May.
Massachusetts State Hospital, 1922-1923, May and June.
State Hospitals, northern United States, 1922-1923, March.

After the maximum the diminution in admissions is in all cases very rapid. The number of admissions fell to the lowest point at the following periods:

New York State Hospitals, September-February.
Dr. Norbury's Sanatorium, October-February.
Massachusetts State Hospital, November-February.
State Hospitals, northern United States, August-February.

Across the Atlantic insanity in London reaches a distinct maximum in May and is low from July to February. Almost identical conditions prevail as to suicide, except that the fall from the high point in June is not so rapid as in the case of insanity, although the minimum is reached earlier, that is, in November and December, instead of from July to February. As to the nervous disorders in continental Europe, Garnier states that general paralysis in Paris follows a seasonal course almost identical with that of insanity in London. Now nervous breakdowns, insanity, suicide, and paralysis, as Norbury shows, are all due

---

* Frank Parsons Norbury: Seasonal Curves in Mental Disorders. *Medical Journal and Record*, vol. 119, 1924.

mainly to the same cause, namely fatigue of the nerves. Such fatigue, he says, is apparently controlled to a large degree by the seasons. But just as the maximum number of deaths lags some weeks or even a month or two after the time when the weather produces the lowest efficiency in factories, so the maximum effect of fatigue of the nerves lags still more, and the greatest number of nervous breakdowns may occur three or four months after the period of least efficiency as measured by daily work.

The universality with which the bodily functions respond to the seasons may be judged from two other recent investigations. In one case Hess* has shown that among infants the phosphates of the blood, which are an essential element for growth, show a pronounced seasonal tide. During the year covered by his observations the percentage of phosphates in the blood stood at 4.34 mg. per cent during June and July, 1921. It may have risen higher during the succeeding months, but no records were kept. In December it had fallen to 3.92 and in March to 3.58. Presumably it would have fallen still lower, but whenever the phosphates fell below 3.75 the children were treated with ultraviolet light, which effectively increases not only the phosphates, but also the calcium and probably other important elements of the blood. It has been found by many authorities that the children's disease known as rickets follows a seasonal course like that of the phosphates and is connected more or less closely with the amount of ultra-violet light. For our present purposes, however, the important point is that the essential phosphates in children show the same general seasonal variation as the death rate, and as mental breakdowns among adults, the lag being perhaps greater than in the death rate but less than in mental collapses.

* Alfred F. Hess: A Seasonal Tide of Blood Phosphate in Infants. *The Journal of the American Medical Association*, December 30, 1922, vol. 79, pp. 2210-2212.

Another case of seasonal fluctuations is discussed by Porter*
of the Harvard Medical School. In coöperation with the Health
Department of Boston, monthly records of the weight of several
thousand of the youngest school children were begun in 1909 and
continued until 1919. Among the boys born in 1905 the average
increase in weight from month to month during the years 1911
to 1918 was as follows:

January to February	+0.18 lbs.	July to August	+0.80 lbs.†
February to March	+0.47	August to September	+0.93†
March to April	+0.22	September to October	+0.96
April to May	—0.16	October to November	+0.61
May to June	+0.05	November to December	+0.63
June to July	+0.50†	December to January	+0.98

In interpreting this table, allowance must be made for clothing. In May the children exchange their winter clothes for those
which are somewhat lighter, while about the end of September
the opposite change takes place. Allowance must also be made
for the long summer vacation with its opportunities for outdoor play which naturally causes the children's weight to increase rapidly. If allowance is made for these two facts the
regularity of the seasonal trend of growth is intensified. From
January onward the children grow slowly; in the spring after
the phosphates have reached the minimum and at the very time
when grown people are most subject to mental breakdowns, the
children practically cease to grow. Not until the summer vacation begins do they recover from the effect of the winter. In the
summer the fourfold advantages of freedom from school, favorable weather, much outdoor life, and a more varied and healthful
diet than at other seasons cause rapid gain in weight. This gain

* W. T. Porter: The Seasonal Variations in the Growth of Boston School
Children. *American Journal of Physiology,* May, 1920.

† The boys were not weighed during the summer vacation. The average
increase in weight from June to September was 2.23 pounds, and this has
been allotted to the summer months in what seem to be reasonable proportions.

seems to be checked somewhat when school begins, but is resumed during the late fall and early winter. In spite of confinement in school, little outdoor play, and a diet relatively poor in vita-mines and other important elements, the children gain weight more rapidly in December than in any other month except perhaps at the end of the summer vacation. The chief favorable factor appears to be the climate, the stimulus of which does not disappear until the advent of really cold weather. The sudden decline in the children's rate of gain during January appears to correspond closely with the drop in the efficiency of factory workers at about the same time, but the effect on growth lasts much longer than does the more direct effect upon activity.

Among the many other instances of seasonal fluctuations in human health, one of the most remarkable is illustrated in Figure 14. The upper line indicates for each month the average

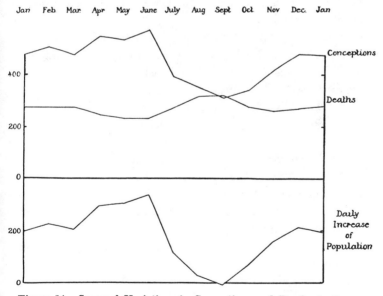

Figure 14. Seasonal Variations in Conceptions and Deaths in Japan, 1901-1910

daily number of conceptions which resulted in the birth of living children in Japan during the ten years from 1901 to 1910. A pronounced maximum in June is followed by a diminution of 46 per cent, which culminates in September at the end of the long, hot, humid summer. Then comes a recovery which is checked but not reversed during the winter, and which resumes its course during the delightful spring weather of April, May, and June. The second curve shows the average number of deaths per day. It is almost exactly the reverse of the curve of conceptions. In other words, during the months when many people are sick and die, the number of conceptions is either very low or else a large number of conceptions result in miscarriages or still-births. The most extraordinary feature of Figure 14 is the fact that the curves of conception and mortality cross one another in September. The same fact is illustrated in the lowest curve, which indicates the excess of conceptions over deaths. In June the conceptions which give rise to living children outnumber the deaths by nearly 2.5 to 1. In September the conceptions are less numerous than the deaths. No seasonal variations in farm work or social customs seem competent to explain more than an insignificant part of the great contrast between May and September. The explanation seems to be that the hot, humid summer saps the vitality of the Japanese, especially the women, so that they are physically unable to reproduce themselves. If the weather which prevails in July and August should prevail throughout the year, the Japanese as a race would apparently diminish in numbers instead of increasing with disquieting rapidity. Under such circumstances natural selection would presumably work with great vigor. A race might arise capable of withstanding the most intense tropical conditions, but it would presumably differ from the present Japanese in many qualities such as energy and initiative.

It would be easy to multiply examples, but space forbids. All sorts of physiological conditions appear to vary from season to season in essentially the same way except that some responses, such as the energy of people in good health, lag only a little after the climatic conditions, others, such as diseases and the rate of reproduction, lag several weeks; and still others, such as the growth of children and the occurrence of nervous breakdowns and suicides, lag still farther. In the case of the mental disturbances the lag is so great that it almost seems as if the onset of stimulating weather after a period of unfavorable weather, had the effect of causing a sudden collapse. In reality the real state of affairs may perhaps be this: the winter months produce an effect like that of driving a horse without rest and as rapidly as possible over a bad road. When a stretch of good road is reached the driver, to whom we may liken the weather, whips the animal to his topmost speed and the tired beast soon breaks down.

Let us now try to analyze the effect of the seasons upon health, and determine the relative part played by temperature, humidity, and variability. A study of about nine million deaths by means of climographs as described in *World Power and Evolution*, leads to the conclusion that the optimum or most favorable condition for human health is an average outside temperature of about 64° F. (18° C.) for day and night, a relative humidity of about 80 per cent, and a fairly high degree of storminess, or at least of variability from day to day. This means a climate in which the midday temperature rises to 70° more or less, while that of the night falls below 60°. With the rise in temperature the noonday relative humidity declines to perhaps 60 per cent, while during the cool night it rises high enough so that dew is precipitated. But a constant succession of clear days is not desirable. There must be occasional rains and variations in temperature, wind, and cloudiness from day to day. Tampa experiences such conditions at the end of February, New Orleans in March, Asheville in May, Atlantic City in early June, Seattle in August, Nantucket and Boston in September, and Portland,

Oregon, in October. At the seasons when people go in largest numbers to many famous health resorts the majority of such places enjoy climatic conditions closely approaching those which are ideal for physical health.

EFFECT OF TEMPERATURE ON HEALTH AND STRENGTH

A Nature of criteria	B No. of persons	C Best temperature	D Effect of change of 10°F. on death rate
1. Deaths, north Italy, 1899-1913	781,000	58°	21%
2. Piece-work, men, two Connecticut factories, 1910-1913	300	59°	—
3. Piece-work, girls, one Connecticut factory, 1911-1913	200	60°	—
4. Deaths, British Columbia, 1914-1916,	8,000	62°	25%
5. Deaths, southeastern United States, 1900-1912	122,000	62°	10%
6. Piece-work, Pittsburgh, 1910-1913,	9,000	63°	—
7. Deaths, southern Italy, 1899-1913	752,000	63°	27%
8. Cigar-making, Cubans, Tampa, Florida, 1913	380	64°	—
9. Deaths, southern France, 1901-1910	838,000	64°	17%
10. Deaths, northeastern United States, 1900-1912	2,500,000	64°	15%
11. Deaths, whites, eastern United States, 1912-1915	921,000	64°	12%
12. Deaths, dry interior of United States, 1900-1912	71,000	64°	15%
13. Deaths, northern France, 1901-1910	1,315,000	65°	17%
14. Deaths, east central United States, 1901-1910	739,000	65°	18%
15. Cigar-making, Cubans, Tampa, Florida, 1913	400	65°	—
16. Deaths, negroes, eastern United States, 1912-1915	167,000	68°	12%
17. Cigar-making, Cubans, Tampa, Florida, 1912	380	68°	—
18. Deaths, California, 1900-1912	142,000	70°	18%

The preceding table illustrates the type of statistical evidence on which is based the conclusion that a mean temperature of 64° F. is the optimum. The experimental evidence will be illustrated later.

Columns A, B, and C explain themselves. Column D indicates the approximate percentage by which a change of 10° F. raises or lowers the death rate when the temperature ranges between 30° and 60°. When all seasons are taken together the net effect of a rise of temperature under such conditions is to lower the death rate, while a fall increases the death rate. This, however, applies only to a rise or fall in which the new condition of temperature endures for some time, as in the change from season to season.

In twelve of the eighteen cases in this table the optimum outside temperature was from 62° to 65° F., and in five cases 64° F. Two of the cases where the optimum falls to 60° or lower were piece-work in Connecticut factories. This may be because such work involves mental as well as physical activity. We have found some evidence that the optimum temperature for mental work is considerably below that for physical work. Naturally an occupation where mental and physical alertness are both needed would be most favored by a temperature between the best temperatures for the body and the mind. As to the low optimum of the north Italians, 58°, I have no explanation. The cause could perhaps be detected by a study of the other elements of the weather such as humidity and wind, or of local diseases such as malaria. It is worth noting, however, that Campani* from an analysis of 24,500 deaths at Milan obtains results closely similar to those here set forth. His results, to be sure, are especially important in respect to variability rather than temperature. He finds that deaths are least numerous just after storms while the wind is blowing. They are most numerous in still air during periods of stagnation and after periods with little change

* A. Campani: *Gazetta degli Ospeddiedelle Cliniche*, Milan.

of temperature. Changes of temperature are beneficial in north Italy just as in America.

Among the three cases of the preceding table where the optimum temperature is above 65°, one represents Cubans of Spanish descent but with a good deal of colored admixture, and a second represents negroes. In the first case life in a tropical climate has presumably raised the optimum temperature somewhat. In the second, although the negroes here dealt with lived largely in the parts of the United States from Maryland northward, they probably still retained an ancestral adaptation to a slightly warmer climate than that which is best for the white race. It is worth noting, however, that the Cubans have spent practically their whole lives where the coolest month averages about 70° and the hottest over 80°, while the ancestors of the negroes have dwelt for untold generations in regions still warmer. Nevertheless the optimum for both groups, 68°, seems to be lower not only than the average temperature of their homes, but than the average for the coolest months in those homes. I might add that for negroes the optimum humidity seems to be a little higher than for white men. This again may be an inheritance from a former environment. Such differences between diverse races suggest that permanent physiological changes take place whereby races become adjusted to diverse climates. The slightness of the differences, however, suggests that such adjustment is very slow and incomplete. An earlier and more fundamental adjustment to climate appears still to be largely dominant, and may represent the climate under which man's chief physical evolution took place.

As for the extreme variation in California, where 70° appears to be the best temperature, I am inclined to think that it is due to accident. The California results depend largely on two cities, San Francisco where the mean monthly temperature never reaches 70°, and Los Angeles where conditions of wind and humidity may account for the favorable conditions during the months of high temperature. It is interesting to note that the average of the six extreme cases in the table is 64.2°, against 63.9° for the twelve medium cases. It must be borne in mind, however, that the best temperature is not the same in dry climates as in moist. We shall return to this later.

The conclusion that an outside mean temperature of about 64° F. is the optimum for Europeans agrees closely with the conclusion now widely accepted that an inside temperature not above 68°, and preferably lower, is the ideal. The effects of deviations from this ideal have been the subject of many careful experimental investigations, among which those carried on by the New York State Ventilation Commission under the chairmanship of Prof. C.-E. A. Winslow are especially notable. The results of these experiments, as set forth in a volume entitled *Ventilation,* are so important that I shall conclude this chapter by quoting several pages which pertain not only to temperature but to other atmospheric conditions.

"The work of previous investigators from Hermans to Leonard Hill has made it abundantly clear that extreme high atmospheric temperatures are highly prejudicial to human health and comfort, and that it is to such temperatures rather than to chemical pollution that the most serious effects of bad air are due. The most important result of our experiments has perhaps been the demonstration that even moderately high temperatures —between 24° and 30° C. (75°-86° F.)—are accompanied by demonstrable harmful results.

"*A. Body Temperature and Circulatory Phenomena.*—We find that the rectal body temperature exhibits a definite relation to the temperature of the atmospheric environment, the body temperature when observed at 8 a. m. during the summer time showing a fairly close parallelism with the average temperature of the outdoor air for the preceding night. In our experimental chamber we found that at 68° F. the rectal temperature and heart rate tended to fall, the Crampton index of vasotone increased, the resistance of the peripheral portion of the circulatory system rose, and the velocity of the blood flow was correspondingly lessened. At 75° F. the rectal temperature, heart rate, and Crampton index changed but slightly. At 86° F. with 80 per cent relative humidity, the rectal temperature and heart rate rose, the Crampton index fell, the peripheral resistance decreased and the velocity of blood flow increased.

"The final average results attained under the three atmospheric conditions were as indicated below.

	20° C. (68° F.) 50 per cent relative humidity	24° C. (75° F.) 50 per cent relative humidity	30° C. (86° F.) 80 per cent relative humidity
Rectal temperature	36.7° C.	37.0° C.	37.4° C.
Heart rate, reclining	66.0	81.0	74.0
Heart rate, standing	77.0	105.0	99.0
Crampton index	60.0	45.0	34.0

"We are not prepared to say whether the maximum heart rates attained at 75° F. as compared with 86° F. are significant; but it is clear that rectal body temperature bears a direct, and the Crampton index an inverse, relation to atmospheric temperature. After vigorous physical work the return of the heart rate to normal was somewhat more prompt at 68° F. than at 75°.

"Systolic and diastolic blood pressure showed no very definite relation to temperature between 68° F. and 86° F.; but at 101° F. blood pressure as well as rectal temperature and heart rate showed marked increases.

"*B. Other Physiological Phenomena.* The rate of respiration was slightly increased by moderately high temperatures, from an average of 17.9 at 68° F. to 19.3 at 75° F. and 19.7 at 86° F.; while at 101° F. the increase was very marked. The dead space of the lungs, the volume of the supplemental air, the alkaline reserve of the blood, the respiratory quotient, the carbohydrate metabolism, the protein metabolism and the total metabolism showed no demonstrable relation to atmospheric temperature under the conditions studied in our experiments.

"*C. Comfort and Mental Efficiency.* So far as the sensations of the subjects are concerned, as evidenced by their votes as to the comfortableness of the experimental chamber, the difference between 68° F. and 75° F. is comparatively slight; but a temperature of 86° F. with 80 per cent relative humidity is distinctly uncomfortable, the average votes falling sharply whenever this condition is reached. This discomfort is not, however, accompanied by any inability to perform mental work; such as naming of colors and opposites, cancellation, mental multiplication, and addition. Subjects urged to maximal performance did equally well under both conditions, the conditions being maintained for four hours a day on five consecutive days, or for eight hours a day on four consecutive days. Longer hours of exposure continued for a prolonged period might of course yield different results. Even when the subject was left free to work or not, as much mental work was accomplished at 75° F. as at 68° F. although in one short experiment a temperature of 86° F. did seem to diminish the inclination to do mental work. At 75° F. typewriting (which involves a certain amount of neuro-muscular activity) seemed to be slightly diminished while performance in mental multiplication was actually increased. In general, however, we have no clear evidence that moderate overheating impairs mental efficiency.

"*D. Influence of Atmospheric Humidity.* Somewhat exhaustive studies of the alleged influence of atmospheric humidity upon mental achievement and comfort have yielded entirely negative results. With relative humidities of 50 per cent and 25 per cent, respectively (all other conditions being the same), there was no significant difference in the votes of the subjects as to their subjective sensations of comfort, no difference in temperature of the air next to the chest, or in pulse rate, and no difference in the performance of a long series of complex neuro-muscular tasks specially designed to test the alleged influence of a dry atmosphere upon nervousness and efficiency. Longer periods of exposure might of course produce effects not detected by us; but it seems clear that exposure to a relative humidity as low as 25 per cent at a temperature of 75° F. for eight hours a day on five days a week does not produce any demonstrable harmful effects.

"*E. Physical Work.* We have demonstrated on the other hand a very marked and significant influence of atmospheric temperature upon the performance of physical work. An increase of room temperature from 68° F. to 75° F. caused a decrease of 15 per cent in the physical work performed by men who were not compelled to maximum effort but were stimulated by a cash bonus; and an increase from 68° F. to 86° F. with 80 per cent relative humidity caused a decrease of 28 per cent in the physical work performed under conditions of maximal effort. The fall at 75° F. was most marked in the afternoon hours when fatigue effects were called into play.

"*F. Susceptibility to Disease.* Finally, we have found very definite evidence of the harmful influence of moderately high atmospheric temperatures, particularly if followed by sudden exposure to low temperatures, in promoting susceptibility to bacterial infection. We find that rabbits maintained at a temperature of 86° F. show a distinctly delayed formation of hemolysins and a slightly reduced agglutinative power, as compared with animals kept at 68° F.; and that rabbits kept at 76°-70° F. and then chilled to 20°-50° or chilled first and kept at 79° F. are much more susceptible to infection than animals kept at 65°-70° F.

"Exhaustive observations of the nasal mucosa of human subjects show that in a warm atmosphere there is an increase, in a cool atmosphere a decrease in the swelling, moisture, and redness of the nasal mucous membranes. Sudden change from a hot to a cold atmosphere, particularly when combined with drafts, produces a moist and distended but anemic condition of the mucosa, presumably highly favorable to bacterial invasion; and workers who have been habitually exposed to high temperatures show a marked excess of atrophic rhinitis. Hot moist air (as in the case of laundry workers) seems to be much more harmful than hot dry air (as in the case of furnace men).

"*II. The Effect of Chemically Vitiated Air upon Physiological State, upon Comfort, and upon Efficiency*

"In parallel with our experiments upon the effect of temperature, we also studied the possible influence of. air, chemically vitiated by human occupancy, but at the same temperature and humidity as fresh air used as a control. The factor of air movement was excluded by the use of room fans to stir up both fresh and stale air so that no difference existed save the very slight changes in oxygen and carbon dioxide and the more obvious changes in odoriferous organic constituents due to respiration and effluvia from the body.

"*A. Physiological Reactions, Comfort, Mental Efficiency, and Resistance to Infection.* In most of the reactions studied in our experiments the influence of chemical vitiation of the atmosphere appeared to be absolutely nil. Temperature and humidity being the same, we compared fresh air containing 5 to 11 parts per 10,000 of carbon dioxide with vitiated air containing 23 to 66 parts and found no difference in body temperature, heart rate, blood pressure, Crampton index, rate of respiration, dead space in the lungs, acidosis of the blood, respiratory quotient, rate of heat production, rate of digestion, and protein metabolism.

"Comfort votes indicated that the subjects were quite unable to distinguish from the standpoint of sensation between the fresh and the stale air conditions; the performance of mental work was quite unaffected by the chemically vitiated atmosphere.

"In a special series of animal experiments, guinea pigs exposed to strong fecal odors for considerable periods failed to exhibit any increased susceptibility to inoculations with foreign bacteria or to injection of diphtheria toxin.

"*B. Physical Work.* In regard to the performance of physical work on the other hand, there appeared to be a distinctly harmful influence of the vitiated air. Temperature and humidity being the same, our subjects performed 9 per cent less work in stale than in fresh air, a difference less marked than that produced by warm as compared with cool air (15 per cent) but apparently significant. When both unfavorable conditions were combined (in warm and stale air) only 77 per cent as much physical work was performed as in cool fresh air.

"*C. Appetite and Growth.* Finally we found a marked influence exerted by stale air upon the appetite for food as determined by serving standard lunches to parallel groups of subjects in stale and fresh air, respectively, but with the same temperature and humidity. In the four different series of experiments which were successfully completed on this basis without the intrusion of interfering factors, the excess of food consumed under fresh air conditions was respectively 4.4, 6.8, 8.6, and 13.6 per cent. Since the probable errors involved in these experiments were relatively very slight it seems evident that the chemical constituents of vitiated air may not only diminish the tendency to do physical work but also the appetite for food.

"This conclusion is strengthened from another direction by demonstration that exposure to strong fecal odors causes a restraining influence upon the rate of growth of guinea pigs during, but not after, the first week of exposure.

"*III. Practical Conclusions in Regard to Ideal Conditions of Ventilation*

"The experiments of the commission have in general confirmed the conclusion of earlier investigators that the first and foremost condition to be avoided in regulating the atmosphere of occupied rooms is an excessively high temperature. We have found that even slight overheating (75° F.) produces the following harmful results:

"(1) A burden upon the heat-regulating system of the body leading to an increased body temperature, an increased heart rate and a marked decrease in general vaso-motor tone as registered by a fall in the Crampton index.

"(2) A slight but definite increase in rate of respiration.

"(3) A considerable decrease in the amount of physical work performed under conditions of equal incentive—a decrease amounting to 15 per cent at 75° F. and to 28 per cent at 86° F.

"(4) A markedly abnormal reaction of the mucous membranes of the nose, leading ultimately to chronic atrophic rhinitis and when followed by chill, producing a moist and distended condition of the membranes calculated to favor bacterial invasion. In animals exposure to high atmospheric temperatures, particularly when followed by chill, diminishes the protective power of the blood and markedly increases general susceptibility to microbic disease.

"For these reasons we believe that the dangers of room overheating are far more serious in their effect upon human health and efficiency than has generally been realized and that every effort should be made to keep the temperature of the schoolroom, the workroom, and the living-room at 68° F. or below.

"With regard to the problem of relative humidity it is obvious that a high moisture content combined with high temperature must always be harmful, since the effect of a humid atmosphere is to decrease the heat loss from the body by evaporation. The specifically harmful influence of unduly low humidity which has been postulated by various writers upon ventilation has, on the other hand, not been apparent in our investigations.

"Our results in regard to the influence of the chemical composition of vitiated air (temperature and humidity effects being excluded) have been generally negative. In two respects, however, our experiments suggest that some chemical constituents of the air of an unventilated room may be objectionable. Such air appears (1) to decrease the appetite of human subjects for food, and (2) to diminish substantially the amount of physical work performed under conditions of equivalent stimulation.

"We may conclude then that the primary condition of good ventilation is the maintenance of a room temperature of 68° F. or below without the production of chilling drafts; but that it is also important, on account of certain subtle but real effects of vitiated air upon appetite and inclination to work, to provide for an air change sufficient to avoid a heavy concentration of effluvia such as was associated in our experiments with a carbon dioxide content of 23 to 66 parts per 10,000."

Except in respect to humidity at moderate and low temperatures this long quotation reënforces the conclusions set forth in this book and in *World Power and Evolution*. It brings out the extreme sensitiveness of human health to atmospheric conditions; it shows that temperature is undoubtedly the most important factor; it adds experimental proof to the widespread opinion that as soon as the temperature rises much above 70°, man's capacity and inclination for physical work decline and his susceptibility to disease increases. If high atmospheric humidity is added to high temperature, as in many tropical countries, the harmful effects are shown to be much accentuated. The fact that the Ventilation Commission detected no effect of high temperature and humidity upon mental work seems to mean merely that when people are subjected to moderately adverse conditions for short periods amounting to less than twenty per cent of the time for a few weeks the physical handicaps do not become pronounced enough to exert a measureable influence upon the mind. It seems only logical to suppose that if the adverse physical conditions arising from high temperature and high humidity were to continue and produce their full effect, the mental powers would ultimately suffer. Thus even this phase of the Commission's work is not inconsistent with the conclusions of this book. In this same connection it is interesting to note that although the Commission made no experiments on the effect of variability, its report contains in several places the specific suggestion that variability may be an important but unconsidered factor. The only point wherein the Commission's conclusions radically differ from my own is in respect to humidity, a subject which we shall consider in the next chapter. In spite of this discrepancy the report of the Ventilation Commission, in its main aspects, confirms the general conclusions of this book as to the relatively depressing effects of tropical climates, regardless of specific diseases; and as to the beneficial effects of relatively cool, variable climates.

CHAPTER VIII

MORTALITY, MOISTURE, AND VARIABILITY

I NASMUCH as there is some disagreement as to the effect of atmospheric moisture upon health, it will be well to examine the evidence carefully. On the basis of the deaths in Paris from 1904 to 1913, Besson,[*] the chief of the "Service Physique et Météorologique" of that city, has come to the conclusion that "on the whole, when the humidity increases, the mortality [from diseases of the respiratory organs] decreases two or three weeks later. . . . If one examines each season separately one finds that this rule fails only in summer when there is no clear result. —According to a widespread opinion humidity acts in an unfavorable manner upon human health. The result announced above cannot fail to surprise many people. One sees that an increase in the number of deaths from diseases of the respiratory organs is on an average preceded not by an increase but by a decrease in the humidity." To test this further, Besson divided the winter months into two groups, one with a mean relative humidity below 86 per cent and averaging 82.4 and the other above 86 per cent and averaging 89.2. He found that the deaths per week when relatively high or low humidity prevailed and in the succeeeding weeks averaged as follows:

	Deaths per week		Advantage of high over low humidity in per cent of low
	Low humidity	High humidity	
The same week	163	161	1.2%
One week after	177	167	6.0%
Two weeks after	185	175	5.7%
Three weeks after	191	183	4.4%
Four weeks after	195	196	0.5%

[*] Louis Besson: Relations Entre les Eléments Météorologiques et la Mortalité. *Annales des Services Techniques d'Hygiene de la Ville de Paris,* 1921.

It seems clear that during these ten years in Paris the drier weeks of winter, even though they were quite moist, were accompanied by a slightly increased death rate from respiratory diseases and were followed in the next two weeks by a death rate about 6 per cent higher than that which followed the moist weeks. Besson ascribes part of this effect to the direction of the wind, but the effect of the wind must be produced largely through humidity and temperature. What part is played by temperature in the results for Paris is not clear.

A study of my own includes temperature as well as humidity, and gives results almost identical with those of Besson. I used the deaths from pneumonia in New York City during the year beginning in April, 1917, and compared them with the weather on the day of death and on the preceding day. All the days with any given temperature were divided into two equal groups on the basis of their relative humidity. Here are the results for the 7200 deaths from lobar pneumonia. Those for broncho-pneumonia were somewhat similar, but much more irregular, presumably because the number of deaths was only a third as great.

AVERAGE DEATHS PER DAY FROM LOBAR PNEUMONIA ON SAME DAY AS GIVEN WEATHER CONDITIONS AND ON SUCCEEDING DAY. (NEW YORK, APRIL 1, 1917, TO MARCH 31, 1918.)

Mean temperature	20° or less	21°-32°	33°-45°	46°-55°	56°-65°	66°-70°	71°-75°	76° or over
Low relative humidity	28.4	26.7	28.7	21.5	19.8	10.5	6.5	6.8
High relative humidity	26.3	25.9	28.0	18.8	15.1	8.2	6.1	6.5
Advantage of moist over dry	8.0%	3.1%	2.5%	14.3%	31.1%	28.0%	6.6%	4.4%

The obvious fact here is that at all temperatures the moist days were better than the dry, as appears in the lower line. The same thing holds true whether we consider the deaths on the same day as the humidity or on the succeeding day. While Besson's data seem to show that dryness renders people susceptible to the initial attack of pneumonia which results in death after a fortnight more or less, the New York data suggest that when the disease nears its crisis a dry day may turn the balance toward death, whereas a moist day turns it toward life. In cold weather, however, this effect is slight, for when the temperature is below 45° the moist days in the preceding table reduce the death rate only 4.5 per cent on an average. At the temperatures of 56° to 70°, on the other hand, an additional grain and a half of water vapor per cubic foot of space, or a difference of roughly 20 per cent in relative humidity, is associated with a diminution of about 30 per cent in the death rate. This is especially important because these are the temperatures at which our houses are kept, or ought to be kept, most of the year. It adds another to the bits of evidence which indicate that for respiratory diseases a dry climate is worse than a moist one. The opposite belief has perhaps become traditional largely because in dry climates people live out of doors. Other things being equal, it is always more healthful to live outdoors rather than indoors.

In this connection it might be added that Besson's conclusions as to the relation of temperature to the death rate from respiratory diseases agree with mine as to the similar relation to deaths from each of the two main types of pneumonia. At temperatures between freezing and 60° F. there is an almost perfectly regular decline in the number of deaths as the temperature rises; then the decline becomes less and less marked until a minimum is reached at about 72°. At higher temperatures a slight increase makes itself apparent, but does not go far because there are only a few days in either Paris or New York when the mean temperature rises much above 75°. Still another investigation, that of Greenburg,* who studied the monthly deaths from pneu-

---

* David Greenburg: Relation of Meteorological Conditions to the Prevalence of Pneumonia. *Journal of American Medical Association*, 1919, p. 252.

monia in New York, Boston, Newark, and Providence, agrees
with the two already cited as to the effect of both temperature
and humidity upon pneumonia. Such close agreement makes it
practically certain that the humidity of the air, as well as the
temperature, is an important element in determining the death
rate from respiratory diseases.

The need of certainty as to the effect of atmospheric humidity
is so great that I shall sum up a number of other examples in the
form of a long table which the non-scientific reader can skip.
Since the effect of humidity varies according to temperature, we
must carefully distinguish between temperatures above and
below the optimum of 64°. At the optimum temperature in every
one of the groups of deaths listed in section A of this table, the
best conditions of health prevailed when the relative humidity
averaged 70 per cent or more for day and night together. At
higher temperatures a relative humidity which averages above
70 per cent (for day and night together) does very decided
harm. At the optimum temperature the effect of humidity seems
to be at a minimum, and a humidity above 70 per cent does little
harm. A lower humidity, although somewhat harmful, has less
effect than at other temperatures, the increase in the death rate
ranging from 5 to 15 per cent, according to the degree of dryness. This may be one reason why the New York Ventilation
Commission found no clear effect of low humidity. Their main
experiments were performed at temperatures close to the optimum. At temperatures below the optimum the effect of humidity
upon the death rate is very clear, much more so than at the optimum. For example, at a temperature of about 40° F. a difference
of only 10 per cent in humidity appears to produce approximately the effect shown in section A of the following table.

EFFECT OF HUMIDITY ON THE DEATH RATE

A. Increase in Monthly Death Rate Accompanying a Decrease of 10 Per Cent
in Relative Humidity at a Temperature of 40° F.

(1) Northeastern United States, 1900-1912 . . . .   0.8%
(2) East central United States, 1900-1912 . . . .   1.3%
(3) Dry interior of the United States, 1900-1912 . . . . .   2.0%
(4) Boston (deaths after operations), 1906-1915 . . . . .   3.1%
(5) Large cities of United States (whites non-contagious dis-
      eases), 1912-1915 . . . .   3.3%
(6) Large cities of United States (negroes, non-contagious dis-
      eases), 1912-1915 . . . .   4.0%
(7) British Columbia, 1914-1916 . . . .   7.0%
(8) Large cities of United States (whites, contagious diseases),
      1912-1915 . . . .   9.5%
(9) Southern France, 1901-1910 . . . .   10.0%
(10) Northern France, 1901-1910 . . . .   11.0%
(11) Northern Italy, 1899-1913 . . . .   12.5%

B. Increase in Monthly Death Rate Accompanying a Decrease of 10 Per Cent
in Relative Humidity at All Temperatures, December-March, 1900-1914

(12) St. Louis 0.5%
(13) New York City 1.2%
(14) San Francisco 3.8%
(15) Baltimore 5.0%
(16) Chicago 7.8%

C. Increase in Daily Death Rate Accompanying a Decrease of 10 Per Cent
in Relative Humidity at All Temperatures During the Influenza
Epidemic in New York City (September-December, 1918)
and Boston (October, 1918-April, 1919)

NOTE. The figures in this section indicate percentages of the average daily
change in the death rate instead of percentages of the actual deaths. Hence
the figures are perhaps five times larger than if they were reckoned as in
A and B.

(17) Onset of influenza, Boston 3.0%
(18) Onset of influenza, New York 6.9%
(19) Deaths from influenza, Boston 11.3%
(20) Deaths from pneumonia, New York 21.6%
(21) Deaths from influenza, New York 26.2%
(22) Deaths from pneumonia, Boston . . . .   —8.3%
(23) Onset of pneumonia, Boston . . . .   —1.8%

This table represents all the available mortality data except those already mentioned and certain others to be given shortly. Practically all the evidence seems to point in the same direction. In 21 out of the 23 sets of data in the table the moister days or months have an advantage over the drier. The negative figures for Boston, Nos. 22 and 23, may be accidental, or may mean that Boston's famous east winds, unlike the moist winds in most places, are really too damp. Taking the table as a whole, an increase of 10 per cent in humidity at low temperatures is correlated with an average decrease of not far from 6 per cent in the death rate.

Let us now turn to quite a different investigation. In Boston I made a study of the number of deaths following operations performed in different kinds of weather. On the basis of about 2300 deaths after operations from 1906-1915 the addition of a grain of moisture per cubic foot of space to the air within doors would have diminished the death rate as follows, provided the inside air thereby acquired the qualities pertaining to the outside air moistened by nature:

APPARENT EFFECT OF ONE GRAIN OF WATER PER CUBIC FOOT
OF AIR IN DIMINISHING THE DEATH RATE AFTER
OPERATIONS IN BOSTON, 1906-1915

A. Mean temperature	Percentage of decrease in deaths for one additional grain of water per cubic foot
30° or less	9.2%
31°-40°	16.5%
41°-50°	8.8%
50°-60°	13.7%
61°-70°	4.6%

Because of the small number of deaths this table is irregular, but the irregularities have little significance. At temperatures below the optimum the death rate was higher after operations

performed in damp weather than after those performed when the
air was dry. This was especially true when the moist weather
continued a day or two after the operations. At high temperatures, however, the effect of humidity is not at all the same as at
low, as appears in Figure 15. Dry conditions are shown toward

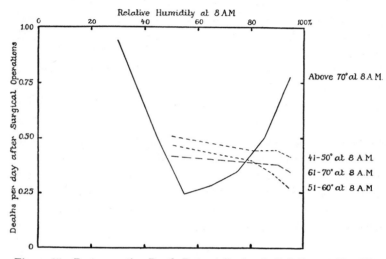

Figure 15.   Post-operative Death Rate at Boston in Relation to Humidity
and Temperature

the left, moist toward the right. The height of the lines shows
the number of deaths per day succeeding operations performed
when various conditions of relative humidity prevailed at 8 a.m.
The broken lines, A, B, and C, indicate the number of deaths in
cool weather. Note their regular decline toward the right. High
humidity was evidently an advantage. Now contrast the dotted
lines with the solid line, A, indicating the effect of humidity
when the temperature at 8 a.m. was above 70°. Under such conditions, far more than in cooler weather, dry air was bad.
Moderately moist air with a relative humidity of 50 to 60 per
cent was better than even the moistest air at lower temperatures.

But note how rapidly the death rate after operations performed in hot weather rises if high humidity is added to great heat. Nevertheless the evil effects of very damp hot days in Boston do not appear to be so bad as those of very dry hot days. This study of surgical operations seems to afford strong evidence of the extreme sensitiveness of the human body to variations in humidity as well as temperature. The fact that it gives a different result at temperatures above and below the optimum tends to establish confidence, for the results at high temperature and high humidity are in perfect accord with common experience and with the experiments carried on by such organizations as the New York Ventilation Commission and the laboratory of the American Society of Heating and Ventilating Engineers at Pittsburgh.

In other countries as well as in France, Italy, and the United States, whence our data have thus far come, the statistics seem usually to indicate that extreme dryness is harmful to health. Other things being equal, the death rates appear generally to be higher in dry regions than in moist. For example, Lucknow and Cairo in dry climates have death rates far higher than Madras and Bombay, which are moist and tropical. Mexico City on its high, cool, but dry plateau, and Johannesburg and Madrid in somewhat similar locations, have exceptionally high death rates in view of their temperature and latitude. In Mexico and India the dry season of March, April, and May has a decidedly higher death rate than the succeeding wet season. This happens in spite of the fact that the temperature in Mexico City during May averages 65° and is almost ideal, while in July it averages a degree or two cooler and is almost equally ideal. In India, on the other hand, both the dry spring and the wet summer are hot, so that the wet season is very muggy. Yet the people heave a sigh of relief when the rains come, for it brings hope of a diminution of disease as well as of good crops.

Let us turn to variability or storminess, another factor which seems to coöperate with temperature and humidity in determining the effect of climate. This factor is especially important because of its bearing on changes of climate and their effect on history. In *World Power and Evolution* the study of climographs based on eight million deaths in France, Italy, and the United States suggests that people's sensitiveness to changes of temperature is almost directly proportional to the uniformity of their climate. For example, in San Francisco a change of only 7° in the mean monthly temperature (50° January, 57° July) is associated with a change in the death rate from 16.4 per cent above normal to 11.7 per cent below, or about 4 per cent for every degree of temperature. At St. Paul and Minneapolis a change from a mean temperature of 12° in January to 67° in June is accompanied by a change in the death rate from 5.5 per cent above normal to 6.9 below, or only 0.23 per cent for each degree of temperature. In the same way a difference of about 9° C. between January and May in Naples is accompanied by a difference of 35.3 per cent in the death rate, whereas in Milan a difference of 16° C. in temperature is associated with a difference of only 26.8 per cent in the death rate. Put in another way this means that the apparent effect of a given change of temperature is 2.3 times as great in Naples of the South as in Milan of the North, and over seventeen times as great in San Francisco with its remarkably uniform climate as in St. Paul and Minneapolis with their severe winters and many storms at all seasons.

In other words, where great changes of weather take place, people become hardened to them. The reality of this hardening is demonstrated again and again by the way in which northerners who move to the tropics lose their power of resistance to even the slightest changes of temperature, but regain it after a few years of renewed residence in a severe climate. The relation between variability and the power to resist disease seems quite clear, but as yet there seems to be no conclusive evidence as to the relative importance of variations from day to day such as accompany storms, and of variations from season to season. The evidence which will now be set forth pertains to changes of temperature from day to day.

In *World Power and Evolution* one of the most important lines of evidence has to do with variations of temperature from one day to the next in New York City. A study of about 400,000 deaths during a period of eight years from 1877 to 1884 shows that when the temperature falls the death rate also falls, while a rise in temperature is regularly accompanied by a corresponding rise in the death rate. This happens not only in summer, when one would expect it, but at all seasons, including even the winter, when one would surely suppose that warm weather would be beneficial. And so it is, in the long run, but the immediate *change* toward warmth is temporarily harmful.

In this investigation, instead of employing the absolute number of deaths, as in previous cases, the *change* in the number of deaths from one day to the next has been used. This is done partly in order that the greater frequency of days with abrupt changes of temperature in winter than in summer may not confuse our results. It is likewise because in dealing with changes of temperature the natural question is whether such changes have any effect in changing the death rate.

The upper part of Figure 16 illustrates another investigation of the same kind in respect to daily deaths from pneumonia in New York. The left-hand side of the diagram indicates the conditions which prevail when the mean temperature of the day of death is higher than that of the preceding day, while the right-hand side indicates that the temperature has fallen. The two solid lines indicate the summer relationships from April to September, A being lobar pneumonia, and B broncho-pneumonia. The dotted lines show winter conditions, C lobar, and D broncho-pneumonia. The significant points about these four curves are as follows:

(1) They are all essentially alike. This suggests that they all conform to some definite physiological law.

(2) Every one of them is low at the two ends and high in the

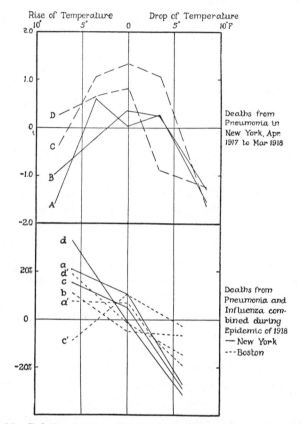

Figure 16.   Relation between Deaths from Pneumonia and Influenza and Interdiurnal Changes of Temperature

middle. This seems to indicate that under normal conditions such as prevailed in the year ending in March, 1918, patients suffering from either form of pneumonia are less likely to die on days when there has been a marked change of temperature in either direction than on days when there is little or no change. This conclusion applies to both summer and winter, but is more true in summer than winter. Hence, for pneumonia patients, a variable climate seems better than one that is uniform. The agreement of the curves with our conclusions derived from factory work and with those of Campani in Italy is noteworthy.

(3) A drop of temperature, on an average, is decidedly more effective than a rise in lowering the death rate. This again agrees with the results of our factory investigation.

The lower part of Figure 16 shows the results of a similar investigation of the influenza epidemic in the latter part of 1918. Because of the enormous variations in the death rate I have here used the percentage rather than the actual number of deaths by which each day's death toll differed from that of the preceding day. Both New York (solid lines) and Boston (dotted lines) have been included. The investigation has been broadened to include (1) the number of cases in which influenza (a) and pneumonia (b) attacked a patient, or at least, in the case of Boston, were reported to have done so, and (2) the number of deaths from influenza (c) and pneumonia (d). The significant points in Figure 16 are as follows:

(1) All the lines except that for deaths from pneumonia in Boston (c') slope in the same way, thus indicating that a rise of temperature is worse than no change, and that a fall is much better than either. This disagrees with the preceding investigation of normal pneumonia, but that may be simply because the lower part of Figure 16 includes only the winter months and not those of the spring when the advent of warm days would be beneficial.

(2) The average changes in the rates for all the conditions shown in the lower part of Figure 16 amount to an increase of 14.1 per cent in the deaths and illnesses on days with a strong rise of temperature and 3.4 per cent on days with little or no change, whereas on days with a strong drop of temperature the

change takes the form of a drop of 18.4 per cent. This means that for all the days when the temperature was markedly different from that of the preceding days, no matter whether it was higher or lower, there was an average drop of 2.2 per cent in the rates of disease and death against a rise of 3.4 per cent on the days of comparatively uniform weather. In other words, here, just as in normal pneumonia, the variable weather had an advantage over the uniform weather. This conclusion is quite contrary to the usual ideas, but it is supported by many other bits of evidence.

One such bit is found in the following data as to the average number of deaths per day in Boston hospitals after surgical operations, when specified changes of temperature took place between the day of an operation and the succeeding day.

Interdiurnal change of temperature	Daily deaths December to February	Daily deaths March to November
Changes of 9° F. or more in either direction .	0.340	0.235
Change of 4° to 8° in either direction . . .	0.336	0.288
Change of 3° or less in either direction . . .	0.326	0.343

The first column suggests that in winter the changes of temperature are too severe in Boston, so that days with only a little change have an advantage of about 4 per cent over the days with a violent change. From March to November, however, the conditions are quite different: operations performed on days with the smallest changes are followed by 41 per cent more deaths than are those performed when the temperature changed most strongly from the day of the operation to the next day. Thus for the year as a whole the variable weather displays a distinct advantage just as in New York.

The net result of our studies thus far is the conclusion that temperature, humidity, and variability all play important parts in determining variations in health and mortality from day to day, month to month, season to season, and year to year. For each climatic factor there appears to be a distinct optimum or most favorable condition, but this varies considerably in response to differences in the other factors. Thus the optimum temperature in dry weather is not the same as in wet, and presumably is not the same in variable weather as in that which is uniform. The high level of the optimum temperature which we found in California may possibly be due in part to the uniformity of the weather. Again, a degree of humidity which is highly favorable at a temperature of 50° F. may work severe harm at 80°, just as a degree of variability which is highly beneficial in warm weather may be too extreme in the midst of a cold winter.

In the rest of this chapter I shall describe certain investigations in which the three climatic factors of temperature, humidity, and variability are analyzed in reference to the influenza epidemic in 1918. The investigation of influenza was carried out by a Committee on the Atmosphere and Man appointed by the National Research Council of the United States. A mathematical method known as partial correlation coefficients was employed. This method has the remarkable quality of picking out and isolating the effect of any one among a number of factors as in an experiment. In the present case the Committee set itself the completion, or at least the extension, of a task already begun by Professor Pearl of Johns Hopkins University. The task was to ascertain whether any environmental conditions were responsible for the great differences in the mortality of the influenza epidemic of 1918 from one city to another. During the ten weeks of the main epidemic Philadelphia, for example, had a death rate from influenza and the resultant pneumonia four times as great as that of Milwaukee, while the rate in Pittsburgh was twice as great as in the neighboring and similar city of Cleveland. Before the committee's investigation was finished the following twenty-two factors had been examined:

A. Factors of human environment (demography).
>   1. Age (proportion of inhabitants of various ages).
>   2. Sex (number of females per hundred males).
>   3. Density of the population (persons per acre within city).
>   4. Rate of growth from 1900 to 1910.

B. Factors of geographical position.
>   5. Distance from Boston, where the epidemic began.
>   6. Longitude.
>   7. Latitude.

C. Physiological factors; normal death rates in 1915, 1916, and 1917 from:
>   8. All causes.
>   9. Pulmonary tuberculosis.
>   10. Organic diseases of the heart.
>   11. Nephritis and acute Bright's disease.
>   12. Typhoid fever.
>   13. Cancer and other malignant tumors.

D. Racial factors.
>   14. Percentage of negroes, 1920.
>   15. Percentage of foreign born, 1920

E. Industrial factor.
>   16. Percentage of population engaged in manufacturing, 1919.

F. Climatic factors.
>   17. Mean temperature for day and night.
>   18. Change of mean temperature from one day to the next.
>   19. Absolute humidity (weight of water vapor per cubic foot of space).
>   20. Relative humidity, or percentage of possible water vapor.
>   21. Weather—a combination of Nos. 17-20.
>   22. Climatic energy as defined in this book.

Directly or indirectly these twenty-two factors embrace most of the conditions which may have been effective in causing people's power of resistance to the epidemic to vary from city to city. Sanitation and medical practice fail to appear in the list because their degree of excellence cannot easily be expressed in figures. But the death rate from typhoid fever is generally supposed to be an unusually good measure of sanitary efficiency, while other death rates are in most places a fairly good index of the excellence of the medical service. Almost the only important field which the factors do not cover is that of variations in the disease-bringing bacteria so far as such variations are due to causes not included in our table. When all these various factors are investigated by means of the most exact and delicate mathematical method yet known, the only one which shows any conclusive causal relation to the destructiveness of this particular epidemic is the weather.

In the work which ultimately led to this conclusion, the Committee on the Atmosphere and Man took the death rate from influenza and pneumonia during the ten weeks succeeding the outbreak of the epidemic in each of thirty-six large cities in the United States. These ten weeks cover the first and, in most places, much the more important outbreak. The committee also obtained data as to the temperature, relative humidity, absolute humidity, and change of temperature from one day to the next. The weather data were tabulated for periods of ten days beginning seventy days before the onset of the epidemic and continuing fifty days thereafter. Previous to the thirtieth day before the epidemic there is evidence of no real relationship between any weather condition and the destructiveness of the influenza. During the thirty days just *before* the onset of the epidemic, however, the temperature and especially the absolute, as distinguished from the relative humidity show a distinct relation to the succeeding death rate. This means that if the weather was warm during the month before the influenza reached a city, the death rate was high; if the amount of moisture in the air was great, the conditions were still worse. At Boston, for example, from the twentieth to the eleventh day before the epidemic the temperature was higher than during the corresponding period in any other cities except New Orleans, New York, and Los Angeles. This was natural, for the epidemic broke out in Boston earlier than elsewhere. In places like St. Paul, Toledo, and Grand Rapids, the cool and fairly dry autumn weather which prevailed for a month before the epidemic apparently gave people a certain degree of stored-up vigor which stood them in good stead and lessened the ravages of the disease. If the temperature was variable, as in Cleveland, Columbus, and Richmond, and especially if it fell during the ten days after the onset of the epidemic, the death rate was lower than where the contrary conditions prevailed. On the other hand, high relative humidity during the ten days before the onset was associated with a relatively high death rate. Cambridge, New Haven, and New Orleans suffered most in this respect. The dampness perhaps made it easy for the bacteria to be transmitted. Droplets of water in the air may act as carriers of the bacteria, or may preserve their virility.

From the tenth to the thirtieth days *after* the onset of the epidemic the virulence of the bacteria was apparently so great that the state of the weather made no difference in the death rate. At any rate there is no evidence that the immediate weather conditions had any effect in overcoming the sudden and sweeping character of the infection. After the thirtieth day, however, there came another change, and the apparent effects of temperature and absolute humidity again rose high. This was the time when in most places the disease reached its maximum and began to decline. At that time cool and moderately dry weather once more was associated with a low death rate. This does not necessarily mean that cold weather is favorable at the time of an epidemic. In fact, quite the contrary may be the case, for very low temperature may be as bad as high. Labrador suffered greatly in the epidemic of 1918.

Having reached the conclusion that atmospheric conditions influenced the severity of the epidemic, the next step was to find a numerical expression for the weather by combining the temperature, humidity, and variability according to their apparent importance. When this had been done, the method of partial correlations was used to compare the weather with the severity of the epidemic and with all the other factors which showed any sign of being important, namely, deaths from tuberculosis, deaths from all causes, deaths from heart diseases, and climatic energy. The weather proved to be the only one whose correlation coefficient was more than four times the probable error, and hence large enough to be significant. The final partial correlation coefficient, when the four other factors named above were held constant, that is eliminated, amounts to 0.57. This is 7.6 times the probable error, which means that there is only one chance in hundreds of millions that we are being misled by accidental agreements between the weather and the death rate.

"Thus," to quote the report of the committee, "the statistical fact is clear. The weather, which means primarily the weather just before the onset of the epidemic and at or just after the climax, is the one factor thus far investigated which shows a clear, pronounced, and persistent relation to the destructiveness of the epidemic.—This does not mean that the weather was in any sense a cause of the epidemic. It is even possible that the weather may be related to the epidemic only indirectly, as is the case with the death rate from heart disease, although no factor capable of producing this result has yet been suggested. Even if the weather is a causal factor in producing variations in the virulence of the epidemic, there is no reason to think that it is the only factor. If the degree of relationship between two variables is proportional to the square of the correlation coefficient, as is sometimes held, the weather, even if it is a direct agent, may be responsible for no more than a third of the variations from city to city. Nor do our high correlation coefficients mean that the weather had anything to do with setting the date of the epidemic or with determining the severity of the 1918 epidemic compared with other epidemics. Neither do they prove anything as to the effect of the weather in other countries, although elsewhere a relationship similar to that found in the United States seems probable. For instance, the British government estimates that in India the death rate from the epidemic was about six times as great as in the United States, while scanty reports from other tropical countries indicate a similar excessive mortality. The one thing which seems clear from the present investigation is that the weather is the one factor whose apparent relationship to the epidemic is not seriously reduced or modified when other conditions are held constant.

"The results of this investigation should be qualified in still another respect. It is not necessary to suppose that other epidemics will show exactly the same relationships as the epidemic of 1918, even though they may be strongly influenced by the weather. In the first place, the epidemic of 1918 was so peculiar in its virulence, its rapid dissemination, its fatality for persons in the prime of life, and in other respects, that it may well have been peculiar in its climatic relationships. In the second place, the epidemic occurred at a season when the approach of cold weather normally exerts a strong stimulating effect in the United States. It is well known that from August to October or even November the death rate normally declines. The epidemic seems to have reflected this condition. Just so far as the weather approached the conditions which prevail at the time when the autumn mortality is lowest, the ravages of the epidemic were checked. At some other, colder season, relatively low temperature and low humidity might be as harmful as high temperature and high humidity appear to have been in September, October, and November, 1918. As a matter of fact, the epidemic of February and March, 1919, shows only a small positive correlation between the monthly death rate and the temperature. Other conditions, perhaps other conditions of weather, were then dominant; or possibly some cities were too cold while others were too warm, a condition which would make the use of correlation coefficients impracticable.

"Finally, even if the weather should prove to be an important factor in causing variations in the virulence of influenza, we still have little evidence as to how its effects are produced. Presumably the weather gives to the human being more or less power of resistance to disease. But it is not improbable that the weather also has an important effect upon the vigor, reproductive rate, or transmission of the disease-bringing bacteria."

CHAPTER IX

HEALTH AND WEATHER

THIS chapter deals with two investigations of the relation between health and the weather. They seem to me the most conclusive evidence yet available along this line because the various weather elements are more clearly separated than in most cases, and because there is no danger of confusing the effects of different seasons. The first investigation pertains to the weather day by day, and is by far the most extensive in which *daily* data have been employed. It represents a coöperative effort carried on by the Committee on the Atmosphere and Man of the National Research Council of the United States and the Metropolitan and New York Life Insurance companies. In addition to the author those most closely concerned in planning the work were Dr. J. Arthur Harris of the Carnegie Institution's laboratory at Cold Spring Harbor and Dr. L. I. Dublin of the Metropolitan Life Insurance Company, but many suggestions were received from others. The results appear to be fairly conclusive as to mean temperature and changes of temperature, but are inconclusive as to relative humidity.

In order to obtain an adequate series of daily mortality data the Committee was obliged to go back to the years 1882 to 1888 in New York City. At that time and for a few years previous the actual day of death was recorded and the facts were summarized by days and published in the annual reports of the New York Board of Health. This highly valuable record has not since been equalled either in New York or elsewhere, so far as I am aware. The Committee used three sets of mortality data:

(1) deaths of children under five years of age; (2) deaths of persons over five years of age; (3) deaths from pneumonia. In order to avoid errors due to the growth of population and the improvement in medical practice, each year was treated as a separate unit, and each category of deaths was reduced to percentages of the average number of deaths per day in that particular year.

For the phase of the investigation here under discussion, the method of partial correlation coefficients was employed. For the layman it may be well to repeat that by this method the effects of different factors can be separated as in an experiment. In order to avoid periods when the temperature is sometimes above and sometimes below the optimum, the months of December to March were chosen. This is also advisable because the effect of the various climatic elements in winter is less understood than their effect in summer. One aim of the work was to eliminate the effect of the seasons and determine whether a mere departure of the weather elements from the normal for any special month has any effect. Accordingly, in preparing the data for the correlation coefficients each month was treated as a separate unit, and the departures of all kinds were reckoned from the averages for December, January, and so forth. This made it possible to ascertain almost beyond question that the weather day by day causes small variations in health which are superposed upon the large seasonal variations.

The Committee used three elements of the weather: (1) the mean daily temperature; (2) the interdiurnal change of mean temperature from one day to the next; and (3) the mean daily relative humidity. In nature the effects of these three elements are inextricably mixed, but the method of partial correlation sorts them out. The results of this sorting appear in Figure 17. The day marked zero is the day on which a given condition of weather occurs. The weather elements on each such day have been compared with the deaths on that day and on each of the

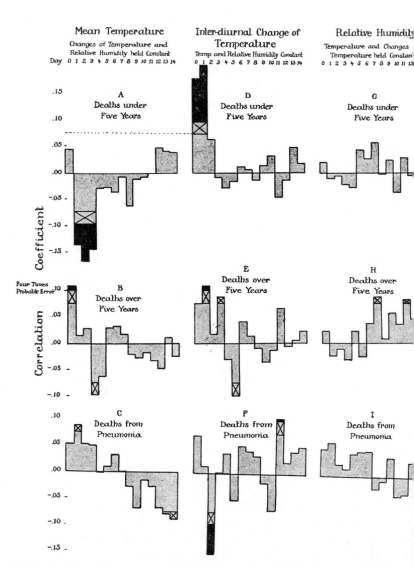

Figure 17. Correlation between Weather Elements and Daily Deaths in New York
December to March, 1882-1886

fourteen succeeding days. The lengths of the bars in Figure 17 indicate the size of the partial correlation coefficients. The three diagrams on the left show the relation between mean temperature and the deaths in our three categories when the interdiurnal change of temperature and the relative humidity are held constant and thus eliminated. The three central diagrams show similar coefficients for changes of temperature when mean temperature and relative humidity are held constant; while the right-hand set apply to relative humidity when mean temperature and changes of temperature are held constant.

Where the bars of Figure 17 lie above the central line, the coefficients are positive; that is, a high condition of the weather, such as high humidity or high temperature, is associated with a high death rate. Where the bars are below this central line the reverse is the case, high temperature, for example, being associated with few deaths. But note that in studying changes of temperature our purpose is not to discover the effect of a small change compared with that of a large change regardless of whether the temperature rises or falls. It is to discover whether the effects of a rise and of a fall are the same or different. Accordingly, the average condition has been counted as that in which there is no change of temperature from one day to the next; a rise of temperature has been given a plus sign and a fall a negative sign. Hence in the central column of Figure 17 a bar above the line means that a high death rate is associated with a rise of temperature, while a bar below the line means that the death rate is high when the temperature falls.

The degree of significance of the bars in Figure 17 may be judged from the shading. Where the bars are lightly shaded their length is less than three times the probable error. This means that they have little or no significance. When the light shading reaches its greatest length, that is, when the coefficient is three times the probable error, there is one chance in twenty-two that a bar of this length would be produced accidentally even if our two sets of figures have no real relation. Suppose for a moment that the death rate and the relative humidity have no real relationship. Nevertheless, with figures the size of those here used we should accidentally get a correlation of 0.075 (three times the probable error) once in every twenty-two correlations. The three diagrams on the right of Figure 17 depict forty-five correlation coefficients. Hence mere chance would be likely to give us two that rise as high as 0.075. What we actually find is two which rise a little above that level.

The areas marked by diagonals indicate values between three and four times the probable error. Now, the likelihood that a coefficient of any particular size will be produced accidentally decreases very rapidly as the coefficients become larger. Thus while there is one chance in twenty-two that a coefficient will be three times the probable error, there is only one chance in one hundred forty-two that it will be four times the probable error. Among the ninety coefficients in the first and second columns of Figure 17 mere chance would not be likely to give more than one coefficient rising to the outer limit of the diagonals, but as a matter of fact we have eleven. If the coefficient is five times the probable error there is only one chance in 1341 that it is due to accident and not to a real relationship; if six times, only one in 19,300; seven, 427,000; eight, 14,700,000. For all practical purposes a coefficient eight times the probable error gives full certainty of a relationship of some sort.

In general we may say that in Figure 17, or any similar diagram, the conditions that *suggest* a relationship are (1) coefficients more than three times the probable error; (2) a considerable series of coefficients all having the same sign and hence all falling either above or below the central line; (3) a series of coefficients which systematically change from high values to low or from positive to negative, or vice versa. The conditions which are generally agreed to amount to practical proof of a relationship are (1) individual coefficients which rise to at least six times

the probable error; (2) several successive coefficients rising to
at least four times the probable error; (3) a considerable series
of coefficients which systematically and persistently change their
values in some orderly sequence such as from three or four times
the probable error on the plus side to an equal value on the
negative side. All three types are represented in Figure 17.

We are now ready to interpret that Figure. Bear in mind that
the length of the columns indicates the *degree of relationship*
between deaths and the weather elements, and has nothing to do
with the actual number of deaths. Remember also that when we
speak of temperature in what follows, we mean temperature after
the effects of relative humidity and interdiurnal changes of temperature have been eliminated by means of partial correlations.
In similar fashion relative humidity and changes of temperature
mean those two factors individually after the other two have
been eliminated. To begin in the upper left-hand corner, the position of the bar for day 0 above the central line in diagram A
indicates that during the 726 days from December to March in
the six years under discussion, high temperature on any particular day tended to be accompanied by a large number of
deaths of children under five years of age on that same day. The
small size of the bar, however, indicates that this relationship
is too small to be considered seriously. On the other hand, the
temperature on any given day had a pronounced relation to the
deaths during the next three days, as appears in the heavy black
shading of days 1 to 3. High temperature was systematically
followed by a low death rate, and low temperature by a high
death rate. Inasmuch as we are dealing with only a single day's
weather, and inasmuch as the largest coefficient (—0.167) is
seven times the probable error, the total effect of the mean temperature for all days on the deaths among children must be
great. On the fourth day after a given temperature, however,
it has practically disappeared.

Diagram B suggests that the reaction of older persons to the outside temperature is similar to that of little children. Curiously enough, however, on the day when a given temperature occurs the effect is stronger than among the children, while on succeeding days the opposite effect is weaker and is somewhat more delayed. Probably the immediate harmful effect of high winter temperature arises from the fact that when the outer air is unusually warm for the season, our houses are likely to be kept too warm, especially on the first day of such warmth. On the other hand, when unusually cold days arrive, many houses which have been too warm become cooler and that is helpful, but soon the fires are pushed and the old condition of hot, stuffy rooms returns.

In diagram C, showing the relationship between the mean temperature and the deaths from pneumonia, the most important feature is the fairly regular decline from a moderately high level on the left to a low level on the right. This apparently means that temperatures which are high for the season tend to cause death among pneumonia patients, but have a good effect in preventing other people from contracting the disease, so that the death rate from pneumonia falls off after about two weeks. Further comment on these first three diagrams is unnecessary. They confirm the results obtained in other ways, and show that the temperature of even a single day plays an appreciable and measurable part in determining the general health of the community.

Look now at the middle column of Figure 17. This depicts the relationship between the death rate and the change of temperature from one day to the next when the mean temperature and the relative humidity are both eliminated by means of partial correlations. Among children less than five years old, as appears at the top, a rise of temperature tends strongly to cause many deaths on the day when it occurs and on the succeeding day, while a drop acts in the opposite fashion. So strong is this effect that the largest partial correlation coefficient (0.202) is 8.4

times the probable error. Inasmuch as there is scarcely one chance in one hundred million that so large a coefficient should be accidentally obtained, we may be practically certain that changes of temperature from one day to the next (regardless of the mean temperature) exert an important effect upon the health of young children. The suddenness with which this effect comes to an end is noteworthy. The portion of diagram D from the third to the fourteenth day is typical of the coefficients obtained when there is no relationship between two sets of phenomena.

Diagram E indicates that among older people changes of temperature from day to day have almost the same effect as among children, a rise being harmful and a drop beneficial. In this case, however, the relationship is not so marked as among the children, the delay is greater, and there is a reaction on the fifth day. Thus the harm done by a rise of temperature, or the good done by a fall, is partly neutralized by effects of the opposite kind a few days later, but the neutralization is only partial, as appears from the greater size of the shaded areas above than below the zero line.

Pneumonia patients (diagram F) present another case where effects of opposite types occur at different times. People suffering from this disease are probably harmed somewhat by a rise of temperature on the very day when it happens, and are similarly helped by a fall, but these effects are too slight to be significant. Two days after a given change of temperature, on the contrary, the pneumonia patients show a distinct benefit if the change has been toward warmer conditions; they are harmed by a change in the opposite direction. This occurs regardless of whether the actual mean temperature is high or low, for that factor has been eliminated by our partial correlations, as has relative humidity. The *changes themselves* appear to be the effective agent. But how about the relatively high positive correlation on the eleventh day in diagram F? There is about one chance in one hundred and fifty that this is due to accident, whereas there is only one in about fifty thousand that the larger coefficient of the second day is accidental. Nevertheless, the positive correlation on the eleventh day may be significant. If so, it presumably means that a rise of temperature is accompanied by conditions favorable to the development of pneumonia, so that an unusually large number of people die about eleven days afterward. Here again we appear to have a curious contradiction between the effect of relatively high temperature and of the *change* toward such a temperature. This contrast appears so constantly and consistently that its reality can scarcely be doubted.

One of the clearest and most convincing features of this investigation of daily changes of temperature is its unequivocal character. In all three diagrams (D, E, and F) the high coefficients are either higher or more numerous than in the corresponding diagrams for mean temperature (A, B, and C). This agrees with several other lines of evidence, such as the sensitiveness of people in monotonous climates, in suggesting that variability of temperature not only from season to season but from day to day may be almost as important as the mean temperature itself. Such slight evidence as is yet available also suggests that variability in other respects such as sunshine, rainfall, moisture, and wind may have an appreciable effect upon health. The emphasis thus given to *variability* as a distinct factor, apart from the conditions which vary, is of much significance in connection with changes of climate and the relation of climate to the distribution of civilization. It confirms the conclusions derived from our study of factories, general death rates, influenza, pneumonia, and operations, and is itself confirmed by strong evidence which is yet to come.

Turning now to relative humidity, we find that diagram G is completely negative. During the years in question the relative humidity of the air had no appreciable effect upon the health of children under five years old in New York. This is true even

when the effects of mean temperature and changes of temperature are eliminated. Among older people (diagram H) the same
is true so far as any immediate effect is concerned, for days 0
to 6 have irregular and insignificant coefficients. From the sixth
to the fourteenth day after any given condition of relative humidity, however, there is a slight but persistent positive correlation every day. On two days this rises to almost four times the
probable error. This suggests that in winter high humidity may
possibly be favorable to the contraction of diseases from which
people die a week or two later. A similar suggestion in respect
to influenza has already been discussed. The pneumonia diagram
(I), however, has a different aspect. None of its coefficients are
large enough to be significant, but the fact that the first seven
are all positive gives a hint that high relative humidity may
have a slightly unfavorable effect upon pneumonia patients.

It is most perplexing to find that different sets of data give
different indications as to the relation between atmospheric
humidity and health. The present investigation with its almost
negative results, but with a slight suggestion that damp air
facilitates the transmission of harmful micro-organisms, agrees
with the experiments of the New York Ventilation Commission,
and with the results obtained by the Committee on the Atmosphere and Man in its work on influenza. Opposed to this are the
results of what seem to be equally reliable investigations pertaining to deaths after operations, to the death rate from pneumonia by months as set forth by Greenburg, to Besson's inquiry
into the weekly death rate in Paris, and to the monthly death
rate where millions of people were studied as described in *World
Power and Evolution*. Moreover, a new investigation, shortly to
be described, points even more strongly toward atmospheric
humidity as an important agent in promoting health. A possible
explanation of this apparent contradiction may be that humidity affects people in two ways, directly through the skin, nerves,
and lungs; and indirectly through minute organisms that bear

disease. The disease-bearing organisms being very short lived, are quickly influenced by variations in atmospheric moisture. Hence a day or two of unusually high relative humidity may be enough to give them an opportunity to produce disease. Man, on the other hand, may be influenced more slowly so that the beneficial effect of moisture upon him becomes apparent only when he is subjected to moderately moist conditions for some time. Thus relatively long oscillations in health may arise through the effect of atmospheric moisture upon man, and short oscillations through the effect upon the bringers of disease, and the two may easily be of opposite character.

We now come to what seems to me a most conclusive study as to the general effect of the weather upon health. In order to gain a comprehensive view of the variations in this effect from place to place and likewise from season to season, I have made a fresh investigation of the deaths each month from 1900 to 1915 in thirty-three cities of the United States. Every city with over 100,000 population in 1910 has been used so far as mortality data are available. Each city and each month of the year has been treated as a separate unit, the Januaries, Februaries, and so on being divided into two equal groups on the basis of each of the following climatic factors:

(1) Mean daily temperature.

(2) Mean daily relative humidity (average of 8 a.m. and 8 p.m.).

(3) Variability or storminess (number of storms whose centers passed within two hundred miles of the given city, allowing double weight to those within one hundred miles).

(4) Wind (total number of miles per month). Five cities only.

Data for the wind were investigated only at New York, Baltimore, Chicago, St. Louis, and San Francisco. The Chicago data are doubtful because the growth of the city appears gradually to have cut off the wind from the Weather Bureau Station and caused an apparent decline in windiness.

An example will show how the data were used. The eight warmest Januaries in New York averaged 6.0° F. warmer than the eight coldest, and had fewer deaths by 0.6 per cent. In February the excess of temperature in the eight warmest months amounted to 6.5° and their death rate was 4.1 per cent less than that of the cooler months. In March the corresponding figures were 6.4° and 9.7 per cent; in April 3.8° and 4.5 per cent; in May, on the contrary, an excess of 3.5° in temperature was accompanied by a death rate 1.5 per cent greater in the warm months than in the cool months, while in July, although the eight warm months averaged only 2.8° above the eight cooler months, the excess in their death rate rose to 14.2 per cent.

The use of this method brings out many interesting facts on which we cannot now dwell. It eliminates entirely all complications due to the seasons, for each month stands by itself and can be compared with any other month. It likewise shows nothing as to the general effect of the *seasonal* variations of either weather or mortality. It merely shows how the *departures of any particular climatic factor from the normal for that particular month* affect the death rate. Boston, for example, is thus found to be benefited in summer by the coolness of its damp east winds; dampness without coolness is the bane of New Haven in summer; New York, by reason partly of its great size and partly of its fine climate, is unusually regular in its responses to the weather, but for some unexplained and possibly accidental reason is averse to storms in midsummer as well as in midwinter. Although Baltimore is hot in summer, it suffers little harm from humidity even when that factor runs high. Baltimore is likewise benefited at all seasons if it has more than its usual allowance of storms, while Boston, being far north, normally has so many storms that it is better off during the less stormy months except in summer. Chicago, on the other hand, is benefited by storms except during

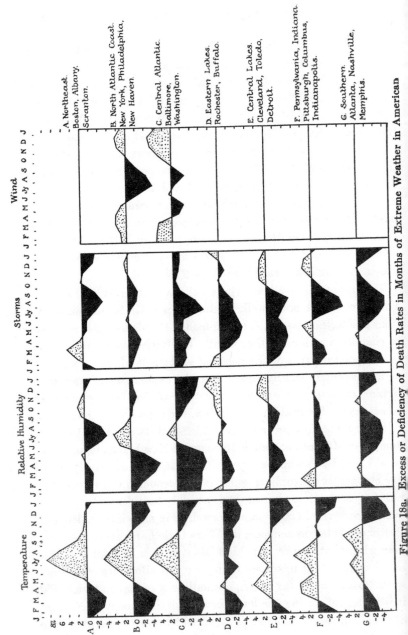

Figure 18a. Excess or Deficiency of Death Rates in Months of Extreme Weather in American

Temperature   Relative Humidity   Storms   Wind

J F M A M J J y A S O N D J J F M A M J J y A S O N D J J F M A M J J y A S O N D J J F M A M J J y A S O N D J

A. Northeast.
Boston, Albany.
Scranton.

B. North Atlantic Coast.
New York, Philadelphia,
New Haven.

C. Central Atlantic.
Baltimore.
Washington.

D. Eastern Lakes.
Rochester, Buffalo.

E. Central Lakes.
Cleveland, Toledo,
Detroit.

F. Pennsylvania, Indiana.
Pittsburgh, Columbus,
Indianapolis.

G. Southern.
Atlanta, Nashville,
Memphis.

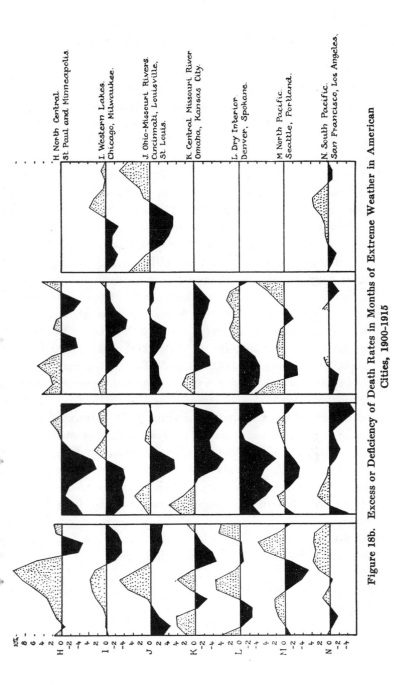

Figure 18b. Excess or Deficiency of Death Rates in Months of Extreme Weather in American Cities, 1900-1915

the coldest months; it is likewise benefited by high relative humidity except in the autumn, but curiously enough it is not benefited by high temperature in winter, perhaps because the warm months experience dry southwest winds, bare ground, and dust. Somewhat the same is true in Pittsburgh and Denver. Cleveland and San Francisco have climates of such a type that departures from the normal produce relatively little effect, whereas in cities like St. Paul and Minneapolis in the north and St. Louis and Cincinnati farther south the effects are much greater. In southerly places like Nashville and Memphis the effect of departures from the normal weather is especially great. Presumably it is still greater in the far south, but Atlanta is the only other southern city for which data are available.

In Figure 18 (a and b) the thirty-three cities have been combined into geographical groups. Each city is weighted according to its population as follows: 100,000 to 200,000 = 1; 200,000 to 400,000 = 2; 400,000 to 1,000,000 = 3; 1,000,000 to 2,000,000 = 4; over 2,000,000 = 5. The curved lines indicate the extent to which the death rates in the eight warmer, moister, or stormier Januaries, Februaries, and so forth, from 1900 to 1915 differed from the corresponding death rates in the eight cooler, drier, and less stormy Januaries and other months. The figures in the scales beside the diagrams indicate departures in percentages of the normals. The normals are the estimated numbers of deaths that each place would have experienced per month in any given year if the number of deaths changed regularly in response to the growth of the city and the improvements in medical practice without regard to weather, seasons, epidemics, and so forth. The method of getting the normals is explained in *World Power and Evolution*. In Figure 18 and the other figures of the same kind all the curves have been smoothed by the formula

$$\frac{a + 2b + c}{4} = b,$$

which is a common way of eliminating the confusing minor irregularities which arise because of the small number of years
for which data are available. In the figure the heavily shaded
areas mean that the months with relatively high temperature,
high humidity, high storminess, or high winds had lower death
rates than the months in which the weather factors stood lower.
For example New York, Philadelphia, and New Haven form a
group of cities lying within a distance of about one hundred and
fifty miles and having similar climates in spite of individual
idiosyncrasies. In the first column of Figure 18, diagram B
shows that in these cities the months of January, February, and
especially March are too cold, for in each month the death rate
in the eight warmest years was from 3 to 5 per cent lower than
in the eight coldest, as appears from the dark shading. In April
the average temperature was about right, for the curve crosses
the zero line; in that month the bad effect produced by weather
that is a little too cool is balanced by the corresponding effect
of high temperature during the same month. As summer advances the warm months begin to have a disadvantage, as is
indicated by the dotted shading. The eight warmer Julies had an
average death rate about 6 per cent greater than that of the
cooler Julies even in the smoothed curve; in the unsmoothed
curve this rises to 10.4. Inasmuch as the eight warmest Julies
averaged only about 3° warmer than the coolest eight, each
degree of excessive temperature raised the death rate more than
3 per cent.

Passing on to the second column of Figure 18 it appears that
high relative humidity is beneficial to New York and its neighbors throughout the winter, and especially in April when an
excess of 6 per cent in relative humidity is accompanied by a
diminution of 6 per cent in the death rate. During the three
summer months the highest humidities do harm, the smoothed
maximum excess of deaths being 4 per cent while the unsmoothed
figures are 5.8 per cent accompanying an excess of 6.4 per cent

in the relative humidity. The duration of the period when humidity is harmful is short. During the autumn and early winter the nearness of the curve to the zero line indicates that it makes little difference whether those months are relatively dry or moist, perhaps because the general conditions of health are so good that people can resist extremes which might harm them at less favorable seasons. Nevertheless, the damper rather than the drier months were the best.

In the third column of Figure 18 the storms of the New York group of cities appear to have had less influence than either the temperature or the humidity. In the late winter and spring the more stormy months were the most healthful; in summer the less stormy ones had a very slight advantage, too small to be significant; the autumn again was like the late winter, while November and December were like June and July. This particular curve happens to be one where the effect of storms is at a minimum. The most important thing about it is that the dark shading is much more extensive than the light, which means that on the whole the stormiest months were times of better health than those that were less stormy.

In the right-hand column of Figure 18, where the effect of the wind is shown, the upper diagram is based only on New York. It is very symmetrical, and indicates that in winter high winds are accompanied by a high death rate, while in summer they are accompanied by a low death rate.

The question at once arises whether the four types of curves in Figure 18 really represent the effect of the individual climatic factors or whether each curve is compounded of the effects of all four factors. For example, are high winds really favorable in summer, or do they merely appear to be so because they are accompanied by low temperature or low humidity or some other favorable condition? The answer is found in Figure 19. The solid curves there are the same as those which we have just been examining for the New York group of cities in Figure 18, but

are plotted on a larger vertical scale. The dotted lines are the
same curves corrected to allow for the other climatic factors.
In making the corrections it was assumed that temperature is
the most important climatic factor, and is followed by humidity,

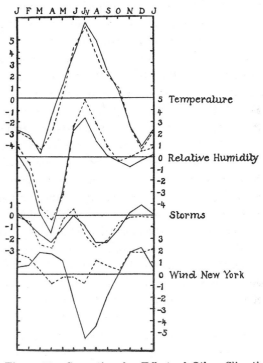

Figure 19.   Correction for Effect of Other Climatic
Factors

storms, and wind in this order, for this is what is commonly
supposed, and it is supported by the investigation here described. The method of making the corrections was as follows:
Knowing how much the temperature and the death rate in the
moister months differed from those in the drier months, it was

easy to determine how much effect a given difference of temperature produces. Since we know the difference in temperature between the moister and drier months, it is possible to make
allowance for this difference month by month and thereby eliminate from the humidity curve the effect of temperature. When
the humidity curve had thus been corrected, the storm curve was
corrected in the same way on the basis not only of the temperature curve but of the corrected humidity curve. Next the humidity curve was corrected to allow for the effect of storms, and the
temperature curve to allow for the effect of both storms and
humidity. Finally, the wind curve was corrected to allow for
variations in all three of the other factors. The result was the
dotted lines in Figure 19. At the top of the diagram the corrected and uncorrected lines showing the effect of temperature
upon the death rate are practically alike. The next pair of lines,
those for humidity, are almost alike, but allowance for the other
factors seems slightly to reduce the importance of humidity in
the spring and raise it in summer. In the third pair of lines the
corrections seem to lower the general level a trifle, thus making
it appear that storms are a bit more beneficial than appeared at
first sight. Finally, the corrected and uncorrected curves for the
winds are completely different. The good effect in summer has
practically disappeared in the corrected curve. If the data for
the wind had been available from New Haven and Philadelphia,
as well as New York, the smoothing of the wind curve might possibly have been still more complete. As things now stand, the corrections seem to indicate that in Figure 18, in spite of minor
details due to other factors, the apparent effects of temperature,
humidity, and storms represent approximately the real effects,
and would represent them still more closely if a larger number
of cities were averaged together as will be done in Figure 20.
As for the wind, it may be that high winds in winter have some
direct effect in raising the death rate, but in summer practically
all of their effect appears to arise from the conditions of tem-

perature, humidity, and storms, or variability which accompany them.

Let us return now to Figure 18. In the left-hand column the groups of cities east of the Mississippi behave almost as one would expect. Except for Rochester and Buffalo (D), which appear to be practically never too hot, all the diagrams are heavily shaded in winter and lightly shaded in summer, thus indicating that the winters from Tennessee and northern Georgia to Minneapolis and Boston are too cold and the summers too hot for the best health. In the center of this area, to be sure, group E (Cleveland, Toledo, and Detroit), group F (Pittsburgh, Columbus, and Indianapolis), and group G (Atlanta, Nashville, and Memphis), show a curious depression in summer, as if the harmful effect of hot weather was somehow inhibited, perhaps because the hot winds are dry so that the bad effects of high humidity are mitigated when the temperature rises. West of the Mississippi all of the groups (K to N) suggest that while hot summer weather is generally bad except on the cool Pacific coast, unusually warm winter weather is also often harmful. This is presumably because the warm months are generally dry; and our monthly data seem to show that dryness is almost always harmful in cold weather.

That this last statement is true seems to be abundantly verified by the second column in Figure 18. Here there is no such equal distribution of heavy and light shading as in the diagrams showing the effect of temperature. On the contrary, almost every individual diagram displays a greater area of heavy shading than of light, and some of the diagrams such as C, G, and K have practically no light shading. In general the amount of heavy shading, that is, the good effect of atmospheric moisture, increases as one goes from the moister and cooler parts of the country to those that are warmer and drier. It reaches a maximum in diagram L for Denver and Spokane, the two cities in our list where the atmospheric moisture is least. This clearly means that in practically all parts of the United States, so far as data are available, and especially in the drier parts, the health of the inhabitants would be materially improved if there were more atmospheric moisture. This clear-cut and apparently unequivocal result agrees with the study of deaths set forth in *World Power and Evolution*, and with the study of the death rate after operations. The contrast between these three lines of evidence, on the one hand, and the results of our investigation of daily deaths in New York, together with the work of the Ventilation Commission, on the other, is the reason for our suggestion that humidity has two diverse and opposed effects. It seems to be beneficial in its direct effect, except at temperatures above the optimum, and harmful in its indirect effect through bacteria.

Much of what has been said of atmospheric moisture is likewise true of storms, as appears in the third column of Figure 18. Notice how largely the heavy shading predominates. Note also that it is scarce in northern groups of cities such as B, H, and M, but increases as one goes southward until in groups G, J, and especially C practically every month of the year shows a lower death rate when storms are relatively abundant than when they are few. Here, just as in the case of humidity, the regions which have few storms, like those which have little atmospheric moisture, give heavily shaded diagrams because an increase in the number of storms is an advantage to health at practically all seasons. But places like Boston, which are exposed not only to many storms but to strong oceanic winds, may get too many storms in the winter.

The question of the effect of storminess is so important that I have prepared Figure 20 to show what happens when stormy periods last several months. The upper diagram in each case shows the conditions when the storms of a given month are compared with the deaths of that month. In the second diagram the months of the sixteen years used in our study have been grouped into halves according to the number of storms not only in the month when deaths occurred, but in the preceding month. In the lower diagram the storminess of three consecutive months has been compared with the deaths in the third month. To begin with Boston, a relatively high degree of storminess lasting only a single month is slightly beneficial in summer and again, curiously enough, in winter, but this may be a mere accident due to the shortness of our record. If the stormy period lasts two months the good effect of storms is much increased. In fact, Boston's health would apparently be distinctly improved if the city could have frequent periods of relatively high storminess lasting two months during the summer, but not during the late winter. The lightly shaded area in the autumn is so small that it may be accidental. If the periods of storminess last three months, however, Boston gets too much of them and the death rate rises markedly. In other words, Boston seems to lie close to the fortunate level where it gets neither too many nor too few storms in the long run, although in the more extreme periods it gets too many, just as in milder periods it does not get enough.

Contrast Boston with Chicago in Figure 20. A single month of more than the average storminess helps Chicago a good deal during all seasons except midwinter. But two successive months of unusual storminess, and especially three, do harm at practically all seasons. In other words, an increase in storminess hurts Chicago more than Boston. Nevertheless, both cities evidently profit greatly by the fact that they have many storms, as appears when they are compared with cities farther south. In New York, for example, Figure 20 shows that increased storminess during periods of more than one month is beneficial in summer but not in winter, while Seattle is benefited by increased and prolonged storminess at practically all seasons. This simply means that New York, with less severe storms than Chicago or Boston, would profit by a mild increase in storminess. On the other hand, Seattle, with far less storminess than the other cities would be better off to have decidedly more.

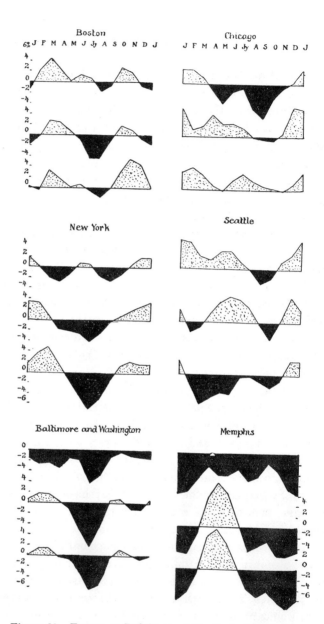

Figure 20. Excess or Deficiency of Deaths in Relation to
Stormy Periods Lasting One Month (Upper Diagrams),
Two Months (Middle), and Three Months (Lower),
1900-1915

At the bottom of Figure 20 Baltimore and Washington, which are treated as a single unit, and Memphis on the Mississippi River in Tennessee present still a third type. At all seasons their stormiest months are the most healthful, for these cities lie toward the southern edge of the storm belt. They have far less stormy weather than New York, Chicago, and Boston, but more than Seattle. During the years under discussion all these places were especially stormy in March. They are benefited by prolonged periods of storminess in summer and autumn, but cannot stand such periods in the winter and spring. Too much emphasis must not be placed on the minor details of any of the diagrams in Figure 20, especially that of a small city like Memphis, for the number of years included in our data is small. Nevertheless, there seems to be little question that storminess has an important effect upon health. In a belt of country extending from New York and Boston westward to Chicago, the beneficial effects of storminess are greatest. To the north of that belt increased storminess appears to have a harmful effect upon health; to the south the present degree of storminess is not enough on an average, and a higher degree regularly causes an improvement in health at most seasons.

In concluding this chapter let us turn to Figure 21. Here the data for all of our thirty-three American cities have been combined into a single diagram. This shows the effect produced by an unusually high condition of any one of the four climatic factors upon the death rate in the whole northern United States together with the Pacific coast. In using such a large area there is great opportunity for opposed conditions in different regions to cancel one another, but this cancellation is far from complete. Before we interpret Figure 21 let us consider for a moment the degree to which the diagrams may be the result of chance. In most such cases it is the custom for mathematicians to compute the probable error by means of a formula. Our method, however, whereby cities are weighted according to their population, and

the departures are reckoned from normals which pertain to the
year instead of the individual months, would cause such computations to take an excessive amount of time. Accordingly I have
calculated the data for four months exactly as in the diagrams,
except that pure chance has been allowed to control the choice
of months for each of the two groups of eight years into which
the sixteen years have been divided. The smoothed results for

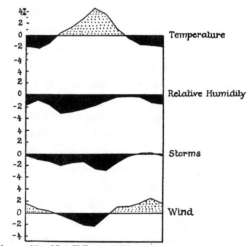

Figure 21.   Net Effect of Weather in the United States

comparison with Figure 21 are January 0.47, April 0.40, July
0.72, October 0.79. In each of the four diagrams the extremes
are four to seven times as great as the accidental variations thus
obtained. This fact, together with the systematic character of
the results, makes it practically certain that we are dealing with
real relationships and not with accidental coincidences.

If this be accepted, our diagrams in Figure 21 show that in
the United States as a whole, excluding the South except in

California, a certain amount of harm is done to health in winter
by low temperature, but not so much as one would expect. The
high temperatures of summer do much more harm. The effects
of the wind seem to be very clear, but are probably due largely
to the conditions of temperature, humidity, and storminess
which accompany the winds. Now for the most surprising fact;
lack of moisture does almost as much harm to the United States
in winter as does low temperature, while in the spring it does
only a little less than does high temperature in summer. At all
seasons the United States as a whole, omitting the parts for
which we have no data, has poorer health in unusually dry
weather than in that which is unusually moist. Almost the same
thing is true of storms. The country is better off when storminess is unusually abundant, except in the late fall. In Figure 21
the average departures from the zero line, regardless of whether
the departures are positive or negative, are as follows: temperature 4.09 per cent, relative humidity 3.96, and storminess 3.60.
These figures seem to represent approximately the relative importance of the three great climatic factors. Thus in spite of
certain puzzling facts as to humidity, the general result of our
study of health in relation to the weather is to confirm the results of our previous study of efficiency in factories. The same
conditions of temperature, humidity, and variability which cause
people to work quickly or slowly in the ordinary affairs of life
seem also to cause their health to be good or poor.

CHAPTER X

THE IDEAL CLIMATE

W E are frequently told that the Riviera or southern California has an ideal climate. Florida lays claim to it in winter, the Alps in summer. Two of the few regions which rarely assert their preëminence in this respect are Boston with its east winds and London with its fogs. Yet in many ways they have a strong claim to high rank. It all depends upon what we mean by "ideal." For rest and recreation a warm, equable climate is doubtless most delightful; for a fishing or climbing trip something quite different is desirable. For most people the really essential thing in life is the ordinary work of every day. Hence, the climate which is best for work may in the long run claim to be the most nearly ideal. But such climates are also the ones that are best for health. Hence they are the ones which people will ultimately choose in the largest numbers. The few disagreeable features at certain seasons are no worse than the shiver at the beginning of a cold plunge.

On the basis of both work and health, the best climate would apparently be one in which the mean temperature rarely falls below the mental optimum of perhaps 38°, or rises above the physical optimum of about 64°. From this point of view the most ideal conditions would seem to be found where the temperature for the year as a whole averages not far from 51°, as at London, Paris, New York, and Peking. In four chief portions of the globe, the winter temperature averages not far from 38°, and that of summer not far from 64°. The first of these is England. At London the thermometer averages 38° in January and

63° in July, while at Liverpool the figures are 39° and 60°. If an average of 51° at all seasons were ideal, southwestern Ireland, with a range of from 45° to 59°, and the Hebrides, from 42° to 55°, would be more ideal than London. On the continent, where the seasonal variation is greater than in Britain, the length of the relatively unfavorable periods with temperatures above 65° and below 38° also increases.

A second region where the temperature conditions approach the ideal is the Pacific coast of the northern United States and southern British Columbia. Seattle, averaging 39° in January and 64° in July has practically the same temperature as London. Southward the seasonal range decreases. San Francisco, averaging 49° in January and 59° in September after the cool summer fogs have passed away, may claim in many ways to be ideal. Still farther south the temperatures of Los Angeles and San Diego, 53° or 54° in January and 69° in August, fluctuate about the physical optimum and would be ideal for physical activity if mean temperature were the only criterion. The mental optimum, however, is lower than the temperature of all except the unusually cold days, and variations from day to day are rare. A short distance inland the Californian climate becomes less favorable than on the coast, for the average in summer at Fresno, for example, is 82°. Even though the heat is mitigated by low humidity, the continuance of such high temperature causes people to feel indisposed to activity.

England and the Pacific coast owe their climatic excellence largely to the fact that ocean winds from the west blow freely over them. Two regions in the southern hemisphere enjoy the same advantage, namely, New Zealand and part of South America, including southern Chili and portions of Patagonia. We are apt to think of these South American regions as sparsely settled places of little importance. This is true, for the present, but it is not because the climate prevents activity. The climate, to be sure, is a drawback, but the harmful feature is not the temperature, but the rainfall. Plants, not man, are the chief sufferers. Unlike our other three regions, this part of the world has a deficient rainfall except in the west, and there high mountains hinder settlement. In spite of this, the few small portions of Patagonia that permit profitable agriculture are making progress and would doubtless do so rapidly if not hampered by remoteness, by the absence of railroads, and by the social handicap of their aborigines.

It must not be inferred that the climates of Patagonia, New Zealand, England, and the Pacific coast of the United States are necessarily ideal. Mean temperature is not the only important factor. Among other things the relative humidity must be considered. The deficiency of moisture in Patagonia not only is disastrous economically, but, to judge from the preceding studies, it lessens man's energy. A similar effect is produced by excess of moisture, and thus harm is done in Ireland and western Scotland, which would otherwise be almost as fortunate as England. New Zealand, the central Pacific coast of North America, and England itself are sometimes unduly damp for long periods, but nevertheless enjoy a relative humidity of not far from 70 per cent much of the time. Other regions, however, such as the eastern United States and central Europe, seem to be more favored in this respect.

The change of temperature from day to day, as we have seen, seems to be as important as relative humidity, and must accordingly be considered more fully. Its effect on human activity seems to be second only to that of the mean temperature of the seasons. The intensity and number of daily changes depend upon two chief factors, first, the range of temperature from the warmest to the coldest part of the year, and second, the number of cyclonic storms. Where the winters are cold and the summers hot, the changes from day to day are also extreme. For instance, in the Dakotas, where the mean temperatures of January and July differ by 60° F., a change of equal magnitude may take place in twenty-four hours. On the other hand, in a place like the Congo, where the difference between the coldest and warmest months is only three or four degrees, the days are correspondingly uniform. The whole matter is illustrated by maps in many physical geographies and in such publications as Bartholomew's *Meteorological Atlas*. The parts of the world where the change of seasons favors a highly advantageous degree of change from one day to the next include most of North America, but omit Florida, the Pacific coast of the United States, and the regions from the Mexican border southward. All of central and eastern Europe is also included except parts of Italy and Greece. A large area in North Africa and a small area in the south of that continent also rank high, as do central Australia and the part of South America which includes central Argentina. Finally, all except the southern parts of Asia lie within the high area. Thus this particular favorable condition occurs not only in many regions whose climate is also good from other standpoints, but in a much larger number whose general climatic conditions are decidedly unfavorable. This is not surprising, for the beneficial effect of pronounced changes of temperature from day to day is often nullified by great heat in summer or extreme cold in winter. Moreover, the seasonal range of temperature forms only one of the two factors which determine the amount of stimulation derived from changes of temperature from day to day.

The other factor is the number of cyclonic storms. By this, as has already been explained, we mean the ordinary storms which produce our changes of weather from day to day in the United States and Europe. Probably the storms are more important than the range of temperature from season to season, for they bring rain, humidity, changes in the winds, and all sorts of stimulating variations. The world's stormiest region, so far as known, includes the Great Lakes region of the United States and southern Canada. Around this center there is an area of great storminess extending southward approximately to Maryland and Kansas, and northwestward through the Dakotas to Alberta. Eastward it includes New England and the Maritime provinces, while northward it quickly disappears. To the south the storminess diminishes gradually, so that Florida has a moderate degree of variability in winter but not in summer. Southern California is the least stormy part of the United States. In Europe the very stormy regions include Britain, most of France, Germany, parts of Scandinavia, and the northern part of Italy, together with western Austria and the Baltic region. In Asia, Japan is the only place where cyclonic storms are at all numerous. The lands of the southern hemisphere generally have few storms. New Zealand is the chief exception, although there they do not cause such great changes of temperature as in America and Europe. The extreme southern tip of South America is likewise stormy, but its storms do not cause much variability. On the contrary, they give rise to a monotony of wind and clouds which is extremely deadening, according to the testimony of those who have lived in such places as Tierra del Fuego or the Falkland Islands. Farther north, in central Argentina, there is a moderate number of storms, comparable to those in the southern United States, and their effect is distinctly favorable.

We are now prepared to estimate the relative stimulating power of the various climates of the world. In England, for example, the mean temperature of the seasons and the degree of storminess are both highly favorable, while the seasonal changes are only moderate. Germany is above medium in temperature, and high in seasonal changes and storminess. In this respect, it resembles the northeastern United States and southern Canada. Japan is similar except that it is somewhat too warm and damp. The coast of British Columbia and of the neighboring states is highly favorable in mean temperature, and medium in storminess and seasonal changes. Around San Francisco, the mean temperature is still better, but both seasonal changes and storms are mild. In compensation for this, however, there are frequent changes of temperature because fogs blow in from the ocean, and are quickly succeeded by the warm, bright weather which generally characterizes the interior. Farther south where the fogs cease, the conditions become less favorable from the point of view of the changes from one day to another, although the mean temperature of the seasons still remains advantageous.

The chief defect of the climate of the California coast is that it is too uniformly stimulating. Perhaps the constant activity which it incites may be a factor in causing nervous disorders. When allowance is made for the fact that California's urban population is relatively smaller than that of states like Massachusetts and New York, insanity appears to be even more prevalent than in those states. Moreover, the cities of the California coast have the highest rate of suicide. In 1922 four California cities led the list in suicides, the number per 100,000 population being: San Diego 47.8, Sacramento 37.9, San Francisco 30.4, and Los Angeles 30.3, against about 15 in the eastern cities. Possibly these facts may be connected with the constant stimulation of the favorable temperature and the lack of relaxation through variations from season to season and from day to day, although other factors must also play a part. The people of California may perhaps be likened to horses which are urged to the limit so that some of them become unduly tired and break down.

In the same way the people of the eastern and central United States are more nervous and active than those of Europe—but not necessarily more efficient—because of still different climatic handicaps. They are alternately stimulated and relaxed by frequent changes from day to day, and in this are like horses that are well driven. In the spring and autumn, however, the combined effect of ideal temperature and highly invigorating daily changes spurs them to an astonishing degree of effort. Then comes the hot summer or the cold winter, either of which is debilitating. People do not diminish their activity at once, especially in the winter. They draw on their nervous energy, and thus exhaust themselves. They are like horses which pull on the bit, and when urged a little break into a run, straining themselves by their extreme speed. Then they are pulled up so suddenly that they are thrown back on their haunches and injured. In Germany somewhat the same conditions prevail, although not to so great an extent. England apparently comes nearer to the ideal than almost any other place. The climate is stimulating at all times, both by reason of abundant storms and because of a moderate seasonal range. It never, however, reaches such extremes as to induce the nervous tension which prevails so largely in the United States.

In strong contrast to these highly favored regions are such places as the center of Asia, where the winters are depressingly cold and the summers unduly hot. The range from season to season is apparently helpful, but its good effects are largely nullified by the infrequency of storms. Day succeeds day with no apparent change. In the desert of Takla-Makan in Chinese Turkestan in the fall of 1905, I found that one of the most surprising features was the way in which winter came upon us unawares. Each morning the thermometer stood a trifle lower than the preceding morning, but there was never any change such as that which we so often experience in America when the first severe frost suddenly comes after a series of days as warm as summer. Frost at last began to prevail at night, but not until we found the water frozen hard in the morning did we realize that winter was upon us. So it goes, month after month, with deadening monotony. Yet when a storm does come the change is often much more extreme than in more oceanic regions. It is frequently so great that its value as a stimulus is much diminished.

Tropical regions suffer even greater disadvantages than do places like the center of Asia. Not only is the temperature unfavorably high, but there are practically no cyclonic storms except in portions where a few hurricanes occur each year.

Thunder storms, to be sure, are abundant, but they rarely bring any important change of temperature. Moreover, the seasonal range from the warmest to the coldest month is generally less than the difference between day and night. Day after day displays no appreciable variation from its predecessor. The uniformity of the climate seems to be more deadly than its heat. Such uniformity, perhaps as much as the high temperature and high humidity, may be one of the most potent causes of the physical debility which affects so many white men within the tropics, and which manifests itself in weaknesses such as drunkenness, immorality, anger, and laziness. Even in tropical highlands the same deadening monotony prevails, although to a less degree than in the lowlands. Such monotony is perhaps the condition which will do most to prevent the white man from living there permanently for generation after generation. His general health may not seem to suffer, but if he works hard he is in great danger of breaking down nervously. The temperature of the highlands may be highly stimulating. There are many places where the mean temperature during every month in the year is within a few degrees of either the physical or mental optimum, or of their average. At Quito in Ecuador, for example, the coldest month, November, averages 54.3° F., and the warmest months, February and September, 55°. Nowhere within the tropics, however, are there any regions which enjoy the physical optimum at one season and the mental optimum at another. If we are justified in associating a high rate of insanity on both the Atlantic and Pacific coasts of the United States with the peculiar climatic conditions, we should expect that white men in tropical regions at high altitudes would suffer still more in the same way, or else would become inert, but no figures seem to be available to determine this point.

We might proceed to discuss scores of ways in which a knowledge of the exact effects of climate may assist in the understanding of historic events, or help in guiding future development. As a step in this direction let us construct a map of the world showing the degree of energy which we should expect among normal Europeans in various regions on the basis of climate. This map will be based entirely on our studies of work among factory operatives and students, and I shall leave the description of it unchanged from the first edition of this book except for the correction of an error. Later we shall test this map of "climatic energy" by one of "climatic health" based on our studies of mortality and disease. Both maps are determined by (1) the mean temperature month by month, (2) the amount of change from one day to another, and (3) the relative humidity. The conditions which prevail at various seasons in the eastern United States duplicate those of almost every portion of the globe. There are hot, dry days like those of the Sahara; hot, damp days like those in the Amazon forests; cold days like those on the great ice-sheet of Greenland, and days of almost every other description. At this point we must make an assumption which cannot be tested until vastly more data have been collected. Let us assume that the continuance of a given condition produces the same effect as its temporary occurrence. For example, in Connecticut our measurements of the effect of days having a mean temperature of 75° are based on occasional days scattered through the summers of several years. Only in rare cases do four or five days of such extreme temperature follow in succession without interruption by more moderate weather. The actual figures show that the first hot day does not greatly diminish people's energy, for the human body is able to resist for a while and to carry the impetus of previous good conditions into the first part of a bad period. After two or three days, however, the heat takes hold on people and makes them inefficient, or even causes some to collapse. If such weather continued for months we should become somewhat accustomed to it, and the period of collapse would be past. Just what the rate of work would then be we cannot yet determine. It would almost certainly be slower than on the first hot day, but it would probably be faster than on the third or fourth. Because of this uncertainty we are obliged for the present to assume that it would be equal to the average of a number of first days and a much smaller number of second, third, fourth, and so on. Having made this assumption, but recognizing that it needs testing, we may go on to construct our map. We must remember that it is not supposed to be a map of the actual energy displayed by the people of various places, but of the energy that we should expect among Europeans if they lived in these places and were influenced as are the people of the eastern United States.

In making such a map it is fortunate that the most important factor is also the one most carefully tabulated by climatologists, and for which our investigations of energy give the most unequivocal results. The mean temperature for every month in the year is given for about 1100 stations in all parts of the world in Hann's *Klimatologie*. The third curve from the top in Figure 10, it will be remembered, shows the average efficiency which would be expected at any given temperature on the basis of the work of factory operatives in Connecticut and New York. A table inserted as an appendix to this book shows the value for each degree of temperature according to the centigrade scale, the maximum being reckoned as 100 at a temperature of 15° C. or 59° F. To determine the effect of mean temperature upon human activity we simply take from Hann the mean temperature of each month, and then from the table in the Appendix, or from the curve in Figure 10, find the corresponding relative efficiency. Then we add the values for all the months. If every month had an average temperature of 59°, with a corresponding relative efficiency of 100, the efficiency for the place in question would be 1200. As a matter of fact this is never reached, but London stands at 1196.6, San Francisco at 1198.6, and Quito in Ecuador at 1198.9. The worst place is Massaua near the southern end of the Red Sea, where the figure is 1070.

The next process in constructing a map of climatic energy is to determine the effect of changes of temperature from day to day. Unfortunately exact statistics are not available in most regions, and we are obliged to employ an approximation. Since changes from day to day depend chiefly upon the seasonal range of temperature and the number of storms, I have combined the two, giving equal weight to each, and giving the two together approximately one half the weight assigned to seasonal temperature. That is, the difference between Quito and Massaua, as stated in the last paragraph, is 128.9. This represents the maximum effect of the seasons, so far as the average temperature is concerned. The maximum effect of the seasons as far as changes from day to day are concerned is reckoned as 30, and the maximum effect of storminess on the same basis is also reckoned as 30. Since highly extreme conditions are not favorable, I have assumed that no seasonal change beyond 30° C., or 54° F., is of value, and also that changes below —7° C. (19.4° F.) or above 23° C. (73.4° F.) are of no value. In other words, if the range from the mean temperature of the coldest to the warmest month is from below —7° C. to above 23° C., it is reckoned as having a value of 30, just as it would be if —7° were the lowest point and 23° the highest. If the range should be from 4° C. to 16° C., it would be reckoned as having a value of 12, while if it were from 20° C. to 28° C., the value would be only 3, because the extremely hot weather above 23° would scarcely be stimulating even if there were slight changes from day to day. In the same way extreme storminess does not produce an effect in proportion to the number of storms. One storm may succeed another so rapidly that the weather ceases to have sufficient variety, and becomes dull and lowering all the time. This is the case at Cape Horn, and also in certain parts of the North American Great Lakes region in winter. Accordingly a storminess of 20 *centers* per year according to Kullmer's scale—which means far more than 20 *storms*—is reckoned as the optimum. Greater storminess is held to have the same stimulating value as 20 centers, while everything lower is counted proportionally. The whole matter is so technical that it cannot be understood without detailed explanations. These are now unnecessary, since the very simple method to be described shortly gives an almost identical but more accurate map.

Humidity has not been considered, because the necessary figures are not available. In most of the cooler parts of the world it would make little difference, although a few unduly damp places like Ireland, or excessively dry regions like Chinese Turkestan would be lower than now appears. The chief difference would be in the warm portions of the world. Agra in northern India, for instance, now has a lower rank than Bombay and Calcutta, but if allowance were made for humidity this would probably be reversed, for Agra is pleasantly dry much of the year. The same reversal would probably occur between dry Khartum and wet Equatorville on the Congo. Arizona and other desert portions of the United States would also make a better appearance than on the present map. It must not be forgotten, however, that our data for New England show that extreme dryness does more harm than extreme humidity. This, however, does not apply to high temperatures. Under such conditions great humidity is undoubtedly most debilitating. Yet even when the air is hot, it may be too dry. In such a place as Death Valley, in summer with the thermometer at 100° to 135° in the shade, it is almost impossible to drink enough water to preserve normal physiological conditions. Even a brief period of physical activity gives rise to much discomfort, and people who stay through the summer are in danger of suffering permanent injury to health.

Our knowledge of the effect of both extreme humidity and extreme dryness is unfortunately still qualitative rather than quantitative. Some day, however, exact figures for all the various climatic elements will be obtainable, and we shall construct a map showing the actual efficiency to be expected in every part of the world. It will be so accurate that the manufacturer, for example, who contemplates establishing a factory, will be able to determine the precise efficiency of labor in the different places which he has in mind, and can put the matter into dollars and cents for comparison with the cost of transportation, raw materials, and other factors.

Meanwhile, our map makes no claim to be more than a first approximation to the truth. Therefore no maps of individual continents are now presented, but merely a map of the world, Figure 22, and of the United States, Figure 34. In preparing these the figures for Hann's stations have been placed on the maps. Then a line has been drawn to include all places falling not more than 25 points below the possible maximum. These are ranked as "very high" and are shaded black. A second line includes places falling from 25 to 50 below the maximum, and the area thus delimited is ranked as "high," and shaded with heavy black lines. The next division, indicated by light lines, is ranked as "medium," and the values range from 50 to 75 below the maximum. The area shaded with thickly scattered spots includes places ranging from 75 to 115 below maximum, and is counted as low. The fifth division, shaded with widely scattered spots, is "very low," and ranges from 115 to 155. Finally, the hot desert areas which fall below 155 are left unshaded, but if humidity were considered they would probably rank as high as the wet parts of the tropics.

Let us now make a similar map on the basis of health. For this purpose I have used the climograph for the eastern United States as given in *World Power and Evolution*. Each column and each horizontal line of the original climograph was plotted, smoothed, and prolonged to the necessary limits of temperature or humidity where these exceeded those actually occurring in the eastern United States. It was assumed that a temperature of 100° F. and a relative humidity of 100 per cent would speedily cause death, and that a mean temperature of 120° for day and night, even with the lowest relative humidity, would ultimately be fatal. From the smoothed data thus prepared the departures of the death rate from the normal under any combination of temperature and humidity can be determined. On the basis of the data previously given it was assumed that the effect of variability is approximately equal to that of temperature or humidity. Since the best degree of variability seems to be about twenty-one storm centers per year according to Kullmer's scale, this number was counted as the optimum. It was assumed that places having this degree of storminess or more have an advantage in health equal to one half of that due to temperature and humidity combined. This combined effect was measured by the difference between the two extremes of Yuma, 7.8 per cent above normal, and San Diego, 6.3 per cent below normal. Although equal importance is thus assigned to temperature, humidity, and storminess in the United States, the importance of storms for the world as a whole amounts to only about one fourth that of the other two combined.

Unfortunately the data as to relative humidity are scanty and lack uniformity. Excellent and uniform figures, however, are available for the United States and India, and less satisfactory figures for Russia and Siberia. These indicate that if complete data were available a map of climatic energy on the basis of the effect of the weather on health in the United States would closely resemble the one based on factory work, as may be seen by comparing Figures 34 and 36. Hence in the absence of good data as to humidity we may for the present use the map of climatic energy based on factory work. (Figure 22.) The similarity of this map to the Indian, Siberian, and American portions of the map based on our data for health and the weather in the United States goes far toward proving its general reliability.

The most noticeable feature of Figure 22 is two large black areas of "very high" energy in the United States and southern

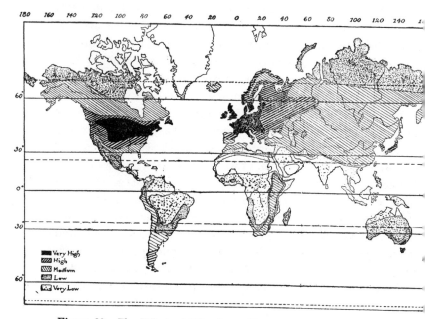

Figure 22. The Effect of Climate on Human Energy as Inferred from
Work in Factories

Canada on the one hand, and in western Europe on the other.
Each is surrounded by a heavily shaded "high" area of large
extent. The remaining high areas, four in number, are surprisingly limited. One lies chiefly in Japan. It is shown as extending
into Korea, but the correctness of this is doubtful, for the
records of storms in that region are imperfect. The second lies
chiefly in New Zealand, but extends into Australia. The records
of storms in this region have been published less fully than in
Europe and America, but the general appearance of the map
seems to be approximately correct. The third of the minor high
areas is located in the southern part of South America. The
records here are very imperfect, and the extent of the high area
is doubtful. The reason for this uncertainty is not only that

reliable records of storms are not abundant, but that the available data do not enable us to determine how much change from day to day is caused by the average storm. The amount of change must be slight compared with that experienced under similar circumstances in North America and Eurasia, because no part of the southern end of South America is far from the ocean. The fourth of the minor high areas lies along the Pacific coast of North America. As already stated, its southern portion owes its character not to storms or seasonal changes, but to frequent breezes blowing in from the ocean. It extends only a short distance inland, and is too narrow to be prominent on the small-scale maps of this volume.

In the far North human energy appears to decline more than would be expected. We know that population is scanty, and civilization low, but we commonly ascribe this to the difficulties of agriculture. Little can be demanded of people who must get a living by hunting and fishing. From the map, however, it appears that even if other circumstances were favorable we should not look for any great achievements. This accords with the slow, inefficient character of the Eskimos, and of the Ostiaks and other inhabitants of northern Siberia. Grenfell in his book on Labrador says that the Eskimo "cannot compare with the Newfoundland white fisherman for perseverance and 'snap.' An Eskimo does not get one fish for the other's ten." This happens even when the Eskimo is in his native habitat, and is doing work to which he has been trained from childhood. Racial inheritance may have much to do with this, but the testimony of white men is that after a long stay in the Arctic they themselves lose ambition and energy.

At the other extreme of climate the regions within thirty degrees of the equator seem to be characterized by essentially the conditions that we should expect. The status of the highlands is striking. A high degree of energy among white men would not be expected permanently in any of them. We are often told that the climate of tropical highlands is as fine as any in the world Not infrequently people are urged to colonize such regions. In book after book we read that there is not the slightest reason why the white man should not live there as well as at home. I do not assert that this may not be possible. In fact, I strongly hope that some day it will come to pass. Nevertheless, our map seems to indicate that previous to any such desirable consummation we must greatly increase our knowledge of how to adapt ourselves to nature, and especially of how to select physical types which are capable of preserving their own health and raising children in spite of the monotonous climate. At present, while the white man may learn to preserve his health in tropical regions, he can scarcely expect to retain the vigor which he displays in the more favored parts of Europe.

The most unexpected feature of the map is the diminution of energy as one proceeds eastward from western Europe to central Asia. In the deserts of Turkestan and Mongolia, and especially in the Tibetan highland, the map should probably show lower conditions than are actually depicted, but as records are not available, the medium shading has been extended across the whole of the unknown area. Before making the studies here described I should have said that a man in Siberia could be as efficient as in far western Russia in the same latitude. Yet the Baltic Provinces are very high in climatic energy, while eastward there is a steady decline until only medium conditions are reached. The reason is readily apparent. In the first place, the Siberian winter is colder and longer than that of the region near the Baltic Sea. More important than this, however, is the decline in storminess as one passes eastward across Russia into Siberia. The cyclonic centers of low pressure, which constitute storms, are either broken up when they approach Asia in winter, or else swing out toward the sea to avoid the great area of high barometric pressure which lies over the continent during the cold season. Hence, during midwinter the far interior is characterized by clear and extremely cold weather, not hard to bear, but steadily benumbing. In the spring and autumn, on the other hand, storms are fairly frequent, and are often of most terrific intensity. The burans, as they are called, are even worse than our western blizzards, which are the same thing under another name. They destroy cattle and horses by the thousand, and human beings often perish within a hundred yards of their houses. Only when the burans are at an end and the milder storms of the late spring and summer prevail does Siberia enjoy a highly stimulating climate.

The conditions just described afford an interesting commentary on the common idea that the plains of Siberia are to be the scene of a wonderful development of European civilization during the next few centuries. I formerly shared this opinion, but have now been obliged to modify it. While this chapter was being written I spoke of this change of opinion to a Russian friend who has come to America for the sake of greater freedom. "Yes," he said, "that is just what the exiles say. I have many friends who are exiles. When they are sent to Siberia they take books with them and expect to do much work in writing and along other lines. Some plan to carry on linguistic studies, and some to make various other kinds of scientific investigations, but they almost never do it. They say that at first they begin to work with great vigor, but after a year or two their energy declines. They have the desire to work, but do not seem able to do so. They attribute this to being so far from home, and to the lack of stimulating contact with civilization. I think there may be more to it than that, for they seem to lose their energy."

Nansen, in his fine book, *Through Siberia: The Land of the Future*, emphasizes this point. He frequently speaks of the slowness and inertia which he encountered. Here per acre the Siberians raise far smaller crops than the Norwegians, and the main reason assigned by Nansen is lack of care, forethought, and energy in cultivating and fertilizing the soil. "There is no hurry

here," he says, "Siberia is still a country that has a super-
abundance of time, as of everything else; they may think themselves lucky for having so far escaped the nervous stress that
we know too well in Europe." He also quotes Rodishev, one of
the most enlightened members of the Duma, who sums up his
impressions of a journey in Siberia with the statement that the
Siberians are "a people without enterprise or initiative." Yet
the vast body of Russians in Siberia consists of pioneers who
voluntarily went to the new country. The mere fact that they
broke old ties and made the hard journey into the wilderness
bespeaks more than the usual degree of energy and initiative.
Similar people in the United States and Canada display great
activity; in Siberia the climate seems to damp their vigor.

All this suggests that the old Russian autocracy accomplished
its purpose more fully than it realized. It not only exiled many
of its most thoughtful and active people, but sent them to a
place where not only do the isolation and hardships diminish
their power, but where nature insidiously accomplishes exactly
the kind of repression that the authorities desire. From the
standpoint of climate, without respect to the many other factors
which may cause quite other results, the relative positions of
Russia and Siberia do not seem likely to change. Both, we may
rightly hope, are destined to advance far beyond their present
position, but while there is reason to think that western Russia
may approach the standard of western Europe, Siberia suffers
from a handicap which may never let her overtake the Baltic
regions on the west of the great northern republic.

Turning to China, we find that the summers are often debili-
tatingly hot, with a steady, damp heat that is apt to be trying.
The winters, on the other hand, are by no means so long as in
Siberia, nor so severe. Yet they are far worse than in western
Europe, and as bad as in any part of the United States. Cold
waves often sweep down from the north, and are so severe that
instead of being stimulants, they are depressing in regions like

Peking. In the south, however, they are beneficial. Everywhere cyclonic storms are rare, so that there is no stimulus of great importance from that source. This is one of the chief reasons why China does not stand high on the energy map. The northern parts of the country are more favored than those in the south or in the far interior, but the difference is not great. Indeed, the uniformity of all parts is surprising. The disadvantages of high temperature in the south are balanced by those of low in the north. If China were part of a smaller continent her nearness to the moderating influence of the sea would help her much more than is now the case. All through the winter she is under the benumbing control of the vast continent to the west, which not only sends out severe cold waves, but prevents the passage of storms. Japan, on the contrary, does not suffer so much in this way. Extremes of temperature are milder than in China, and stimulating storms are frequent. Her greatest drawback is the long period of hot, damp weather in summer. Nevertheless she stands high. Here we must bring our review of the map of climatic energy to a close. We shall come back to it again when we have studied the distribution of civilization.

CHAPTER XI

THE DISTRIBUTION OF CIVILIZATION

DOUBTLESS the reader has already noticed the striking resemblance between the distribution of climatic energy and of civilization. Look again at Figure 22 and see how the black areas agree with the places of highest culture. In view of this it seems advisable to construct a map of civilization to serve as a standard of reference. Only two methods appear feasible. One is by statistics; the other on the basis of opinion. Both present grave difficulties. The statistical method will ultimately prove far the better, but it may not be practicable for centuries. For a fair estimate of the position of a country we need accurate statistics of education, morality, industry, inventions, scientific and artistic skill, wealth, pauperism, charity, crime, and many other aspects of human life which will readily suggest themselves. No reliable figures for many of these things have ever been gathered in any part of the world; no statistics for any of them have ever been gathered in many countries. Statistically it is almost impossible to compare Afghanistan, for example, with Kamchatka. Even where accurate statistics are available, the methods of compiling them are often so diverse as to make comparisons misleading. We may know exactly how many people are arrested and convicted for theft in half a dozen countries, but in one country the police may be so inefficient that few criminals are apprehended, while in another practically every thief may be caught. Thus the better may easily appear the worse. The only way to use statistics at present seems to be as a check upon the other method. We can select some country so extensive that its various parts differ decidedly, but sufficiently homogeneous so that the figures for all portions are comparable. Since the United States meets these conditions better than any other country we shall examine its statistics in several cases, and shall use them as a test of a map prepared on the basis of opinion.

For a map of the civilization of the entire world we must rely on the opinion of well-informed persons, but we shall find that this agrees closely with the indications of statistics. The value of a map based on personal opinion depends partly on our definition of civilization and partly on our confidence in the judgment of the persons in question. Even the best and broadest experience does not eliminate personal or racial bias. Therefore, the only safe course is to obtain the opinions of many people belonging to different races and ruled by different ideals. Accordingly, in the autumn of 1913, I asked over two hundred people in twenty-seven countries to help in preparing a map. Most fortunately this was before the great war broke out. Good feeling prevailed everywhere, and among men of sound judgment there was perhaps as little racial prejudice as at any time during the course of history. This is especially important because similar conditions may not prevail again for years.

The persons whose assistance was asked were selected for various reasons. The larger number were geographers whose first duty is to know all parts of the world. Ethnologists in considerable numbers were included for the same reason, but they responded less freely than the geographers. Historians, diplomats, colonial officials, travelers, missionaries, editors, educators, and business men were all included. The only criterion was that each person should possess an extensive knowledge of the world through personal knowledge, or, in a few cases, through reading. Some were selected because of knowledge of special regions not well known to most people and only reached by extensive travel. To all these many kinds of people, numbering 213 in all, I sent the following letter:

"May I ask your coöperation in the preparation of a map showing the distribution of the higher elements of civilization throughout the world? My purpose is to prepare a map which shall show the distribution of those characteristics which are generally recognized as of the highest value. I mean by this the power of initiative, the capacity for formulating new ideas and for carrying them into effect, the power of self-control, high standards of honesty and morality, the power to lead and to control other races, the capacity for disseminating ideas, and other similar qualities which will readily suggest themselves. These qualities find expression in high ideals, respect for law, inventiveness, ability to develop philosophical systems, stability and honesty of government, a highly developed system of education, the capacity to dominate the less civilized parts of the world, and the ability to carry out far-reaching enterprises covering long periods of time and great areas of the earth's surface.

"In preparing such a map it is evident that statistics may afford much assistance, but they need to be supplemented. They touch only upon material things in most cases, and none are available for a large part of the world. Therefore, our best resource is the personal opinion of competent judges. Accordingly, I am asking a hundred geographers, anthropologists, and other persons of wide knowledge, whether they are willing to take the time to divide the regions indicated in the accompanying list into ten groups according to the criteria mentioned above. Group 10 will include regions of the very highest character, that is, those where the greatest number of valuable qualities are found in high degree. Group 1 will include those which are lowest in these respects. On the basis of this grouping I shall determine the average position of each region and shall prepare a map accordingly.

"The purpose of such a map is threefold. In the first place, it will prove intrinsically interesting to a large number of people, and is likely to arouse considerable discussion. In the second place, in all geographical, historical, sociological, and economic discussions it seems to me that we need a clearer, stronger emphasis upon human character, that is, upon the mental and moral qualities which dominate the civilization of the various nations. If this is so, it is highly important, in the third place, that we should determine much more fully than has yet been the case how far various moral and mental qualities are influenced by physical environment, race, historical development, biological variations, and other causes. In order to determine these things we need a map which shall serve at least approximately as a standard of reference. In discussing the influence of such things as racial character, differences of religion, social institutions, modern means of communication, the form of the land, the relation of land and sea, variations of climate and the like, we shall be able to gain much light by comparing their distribution with that of human character as it now exists according to a consensus of expert opinion.

"The matter can best be illustrated by outlining a specific purpose to which I mean at once to apply the proposed map. [Here follows a brief description of the map of human energy on the basis of factory work.]

"I recognize that those to whom this letter is sent will say at once that they are not sufficiently familiar with all parts of the world, and that they have no means of distinguishing between different parts of China, for example, or between the different portions of equatorial Africa. This is certainly true, but it must be remembered that the classification is very rough. It is only desired that the one hundred and eighty-five names on the enclosed list be divided into ten groups, no group to contain less than fifteen names or more than twenty-one, and each preferably to contain eighteen or nineteen. It may not be easy to determine whether all of the divisions of France, for example, fall in the first group, but it is perfectly evident that none of them will fall in the fifth or lower. The chief thing is to place them as nearly as possible in their proper group according to one's own personal opinion. A given region may properly fall in the fifth group, but the purpose of this classification will not be defeated if it is placed in either the fourth, fifth, or sixth, for when the opinions of one hundred persons are averaged, individual idiosyncrasies will disappear. In view of the varying degrees to which each individual is familiar with the different regions of the world, I should be glad if each contributor would underline the names of regions with which he is well acquainted either by travel or reading, and would place question marks after the names of regions as to which his knowledge is especially deficient. Names not underlined or questioned will be considered as intermediate. The three grades of familiarity thus indicated will be weighted in the ratios of 3, 2, 1. [This has not been done, partly because only about half of the contributors made the division into three grades, and partly because the final results are not appreciably changed by using the unweighted values.] The grade of the various regions should be indicated by underlining or questioning the names upon the small slips mentioned below, but may be done upon the accompanying list if that is more convenient, but in that case please be sure to return the list. For convenience of classification I enclose slips containing the names of the different divisions. These may be spread out upon a table and arranged in ten columns and shifted from column to column until an approximately satisfactory arrangement is reached. When thus arranged those of each column should be placed in the corresponding envelope of the ten here enclosed, and all may be mailed in the large addressed envelope. Envelope ten is for the highest group, and one for the lowest.

"In making the classification, one or two points need to be borne in mind. In the first place, the past should not be considered: Greece, for example, should be placed in the group where its condition during the past one hundred years would place it without reference to its ancient greatness. In the second place, if two races inhabit a given region, both must be considered, and the rank of the region must depend upon the average of the two, giving each one a weight proportional to the number of people. For instance, in a state such as Georgia where nearly half the people are negroes, they must receive half the weight. Still another point is that the rank of a country can often be determined by considering the position which its people take when they migrate elsewhere. For instance, the position of Syrians as compared with Germans when they migrate to England or the United States is a fair criterion as to the relative merits of the two races. After the first generation, however, this should not be applied, for the younger generation owes much of what it is to the new country. A final point concerns countries which are poor in natural resources, or which are not located in the main centers of the world's activity, but which are nevertheless of high character. For example, so far as importance in the affairs of the world is concerned, England vastly outranks Scotland. Nevertheless, our estimate of the greatness of England owes much to the large number of Scotchmen who have gone out to build up the British Empire. Therefore, in estimating the relative merits of Scotland and England, the matter of size or even of commercial importance should receive relatively little consideration, whereas the character and ability of the people as rulers, merchants, scientists, writers, and men of all sorts should have a predominating weight.

"In publishing the final results I should be glad if I might print the names of those who have contributed, but of course this must be as each individual may choose. The individual lists will not be published, and will be treated as confidential. I judge that other contributors will feel, as I do, that their classifications are of necessity so imperfect that they do not care to distribute them to the world at large. Hence, while the list of contributors will be published unless the contributors prefer otherwise, their individual opinions will be withheld. I hope, however, to publish a list showing the average rank of each country and the range of opinion between those who put it highest and those who put it lowest. Inasmuch as the plan here outlined depends upon the coöperation of many contributors, no single individual can in any respect be held responsible for features of the final map which do not meet his approval.

*"In addition to the general list of divisions, I enclose a set of cards bearing the names of the states of the United States, and of the provinces of Canada. Would you be willing to arrange these in groups and place them in the proper envelopes, employing the same method as for the larger divisions? Group 1 will be the states or provinces which are least progressive, or least influential, so far as the general character of their citizens is concerned, and Group 6 the highest. Each group should contain about ten names. The object of this you see is to make a map of the United States and southern Canada on the same basis as that of the world, but on a more minute scale.

"The rough grouping here suggested ought not to take more than a few hours' time. Many days, to be sure, might be devoted to it, but the added accuracy thus gained would not be sufficient to make it worth while. If you can give the necessary time at your earliest convenience I shall be most grateful. If you cannot, would you be willing to return the list, the slips, and the envelopes in order that I may ask someone else to do it in your place? Whether you contribute or not, I shall take pleasure in sending you copies of the final results. Trusting that I may hear from you soon, I am

"Very truly yours,

"ELLSWORTH HUNTINGTON."

---

* This paragraph was included only in the letters sent to Americans and to one or two Europeans especially familiar with America.

Replies were received from 137 persons, while others sent copies of their publications, so that an answer of some sort came from about three fourths of those addressed. The majority of the remaining fourth were foreigners to whom a six-page letter in English might appear formidable. About 90 per cent of the English and Americans sent replies, which is a very large proportion as such things go. I am convinced that the rest failed to answer chiefly because the classification required more time and was more difficult than I at first realized. The fact that classifications continued to be received for an entire year indicates that many meant to answer, but put my letter aside for a more convenient season which never arrived. A third of those who replied, fifty-four to be exact, actually made classifications, and all but two or three conformed so closely to the general plan that it has been possible to use them. The names of the contributors are given in the Appendix. I take this opportunity to express the warmest appreciation of their kindness in coöperating so cordially. Not only their classifications, but their letters were of the highest value and in many cases contained suggestions which have been of great assistance in preparing this volume. The same is true of many letters from persons who did not contribute, but who took pains to explain their reasons and to suggest ways in which my plan might be improved. Except where direct quotations are employed I have not attempted to acknowledge my indebtedness for various ideas which distinctly modify the tenor of these pages. This is partly because the same thought was often expressed by several persons, and partly because in many cases I cannot tell from which of several letters an idea was derived. Except in two instances I have also refrained from mentioning names, because where so many have contributed materials of great value, it might seem invidious to mention some and not others. Therefore, I can merely express my gratitude to all concerned. The net result of this attempt at scientific coöperation among men of many races and tongues leaves a strong impression of the spirit of fellowship and friendly help-fulness among men of wide interests in all portions of the world.

The countries represented in the final classification and the number of contributors are as follows: Australia 1, Canada 1, Norway 1, Sweden 1, Netherlands 1, Russia 1, Spain 1, Portugal 1, France 2, Italy 2, Japan 3, China 3, Germany 5, Great Britain 7, and the United States 24. The number of Chinese and Japanese is particularly gratifying. The ratio between the number of contributors and the number to whom letters were sent is higher among them than among any other main group except the Americans, as may be seen in the Appendix. It is to be re-gretted that no one from India or South America coöperated, and only one, a Russian, from the European countries east of Germany.

The difficulty of making the classification is considerable. Several contributors spoke of spending an entire day upon it, or of taking out the slips time after time to arrange them more satisfactorily. Some said that they spent two entire days upon it. All seemed impressed by the way in which a systematic classi-fication of this kind brings out the weak spots in a man's knowl-edge. For instance, here is the way in which one contributor expressed himself:

"One appreciates what a big world this is and how little one knows about it when he attempts such a task as you have set. It is a most excellent means of taking the conceit out of one."

Another puts it in this way:

"I must confess that it is *the* most difficult and one of the most humiliating games I have ever tried to play! I always knew I was a fraud as a President of a Geographical Society, but I never knew before how great was my deception! The greatest difficulty I found lay through my ignorance of the proportion of the different races inhabiting a district."

An interesting feature of the letters was the diversity of opinion as to the advisability of any such classification of countries. To take the adverse opinions first, one of my best friends, an American geographer, put the matter very strongly:

"I am complying with your request for a sorting of the slips you sent me. It is a very bad plan, and not, I think, of value. Indeed I am not sure that I would have done it for anyone else than you."

I am glad to say that later he expressed a much less severe opinion. Another geographer, a Teutonic European, speaks most cordially in part of his letter, but comes out bluntly in opposition to this particular plan:

"I am wholly unable to take part in this work. I take your scheme as a failure . . . I guess you are here, like some other Americans, under the influence of a too systematizing spirit. It seems to me impossible to classify mankind by this simple method."

Still a third, an American anthropologist, is equally uncompromising:

"Speaking frankly I do not conceive that the method you suggest is possible of scientific results. One must choose between statistics which are definite and mere judgments which are general. To apply the geographic method to a compound of statistics and loose generalization may be productive of grave error."

And a fourth, also an American anthropologist, expresses himself as follows:

"It has been my endeavor, in my anthropological studies, to follow the same principles that are laid down for natural sciences; and the first condition of progress is therefore to eliminate the element of subjective value; not that I wish to deny that there are values, but it seems to me necessary to eliminate the peculiar combination of the development of cultural forms and the intrusion of the idea of our estimate of their value, which has nothing to do with these forms. It seems to my mind that in doing so these obtain subjective values, which in themselves may be the subject of interesting studies, but which do not give any answer to the question that you are trying to solve."

Another anthropologist, this time a European, at the end of a particularly long and suggestive letter, expresses himself thus:

"Taking all that I have written here into consideration, I think that if we were going to grind all the different regions of your long list in the same statistical mill, and to try to compute an average, a highly improbable and fantastic result would be obtained. For my own edification I put some of your criteria to the test, though in a different way. I drew up a list of twelve characteristics of the 'highest value,' in which I included sense for beauty in literature (*belles lettres*) and a few others, and then distributed them to eleven different regions of the globe. My familiarity with those regions by a long sojourn or travel *and* reading, covers a lifetime. To each characteristic for each region I assigned a number, from 1 to 10. I then added the different values or points to try to find a ratio, which might be called 'index of civilization.' I give it *valeat quantum valere potest*."

The table possesses so many points of interest that it is inserted below. At the end I have added a column showing the rank of the various regions according to all the contributors, as computed according to the system presently to be outlined. If the plan embodied in this table could be carried out on a large scale, it would undoubtedly be better than mine. The difficulty is that it requires a vast amount of work and a degree of familiarity with the various peoples of the earth which is found only among a few exceptional students who can almost be counted on the fingers. In course of time we may perhaps hope for a map based on some such minute study of human nature. Yet when it is before us, there is every reason to think that its general features, with which alone we are concerned in this book, will be almost identical with those of the map which we shall shortly consider. The reasons for this will be given later.

One important point stands out in this table. I have given too little weight to the æsthetic side of human nature. In framing a definition of civilization I consciously thought of art in all its forms, but it seemed as if this were included in "the capacity for formulating new ideas and for carrying them into effect," just as science is meant to be included. Moreover, the course of history seems to show that every nation which rises high in other respects sooner or later experiences a period of high development in art, architecture, literature, and science. Nevertheless, these things should have received more specific recognition in my definition.

To turn now to the other side of the question, those who believe in the utility of the plan presented in my letter naturally do not feel the necessity of stating their reasons. Nevertheless, a considerable number take pains to express approval. For instance, a widely traveled Englishman thinks that "there are tremendous possibilities in all such attempts." An American, who is familiar with most of the countries of the world, says: "Permit me to say how heartily I thank you for engaging in this enterprise. Despite any and all sympathetic or hostile criticism of such a work, or the cheap and hasty or really valuable appraisal likely to be made, such a scheme will be invaluable to all students of human progress."

Another American, who has spent a large part of his life in the Orient, is of the opinion that: "It is a fascinating and very significant set of standards by which you have asked us to group

DISTRIBUTION OF SOME CHARACTERISTICS OF HIGHEST CIVILIZATION
BY H. TEN KATE

Regions	Power of initiative	Inventiveness and capacity for formulating new ideas	Ability to develop philosophical systems	Ability to carry out far-reaching enterprises covering long periods and areas	Power to lead and control other races	Highly developed system of education	Application of principles of hygiene	Standard of honesty and morality	Degree of personal safety	Sense for beauty in art (architecture included)	Sense for beauty in literature (belles lettres)	Sense for beauty in nature (scenery, flowers, etc.)	Average showing index of civilization	Rank according to 50 contributors
Great Britain Nos. 1 and 4*	10	7	6	10	9	5	7	7	7	4	8	9	89	100
Northern France No. 17	6	8	8	5	3	7	5	2	4	10	10	4	72	99
Netherlands Nos. 6 and 10	3	5	3	2	5	9	7	8	10	10	10	10	82	96
Germany Nos. 2 and 3	8	8	9	6	2	8	8	6	9	9	9	10	92	99
Switzerland No. 7	3	2	2	1	1	8	8	7	9	2	6	4	52	97
Italy Nos. 18 and 21	3	4	3	2	3	4	3	2	5	8	7	2	44	89
United States Nos. 47 and 48	10	9	2	10	1	7	8	6	7	2	6	2	72	100
Mexico Nos. 70, 71 and 74	1	1	1	1	1	1	2	1	1	1	4	1	17	54
Argentina Nos. 164 and 167	2	1	1	1	1	2	4	1	6	2	6	1	28	63
Paraguay No. 169	1	1	1	1	1	1	1	1	7	1	2	1	19	48
Japan No. 77	3	1	2	3	1	5	4	1	7	7	6	5	45	82

*The numbers refer to the detailed list of countries in the Appendix.

the regions. From the time I started after lunch until now, late in the evening, I have done nothing else. But there is more than fascination. There is a very deep issue involved, and I am glad to share in the construction of the map and the charts you are aiming to devise."

One of the fairest estimates of both the advantages and disadvantages of the plan is contained in a letter from "Ambassador" Bryce, as we Americans still like to call him. With his permission I quote it in full.

"Your idea is ingenious and interesting, and I should much like to see how it works out, though it seems to me, on first impressions, that the various factors involved are so many and so complex that the visual presentation you contemplate would need an amount of comment and explanation which would require something like a treatise to accompany the map. You give one instance in the case of the state of Georgia. Another might be found in the other Georgia, south of the Caucasus, where besides the native Georgians (Karthli) there are Armenians, Tartars, and Russians, not to speak of minor races. Or take Japan. If one were to think of the educated and ruling Japanese and the sort of civilization they have created one would make a rating quite inapplicable to the ordinary Japanese. Without therefore doubting that by means of a map presentation very interesting results may be obtained, I should think that some modifications would be needed. 'Efficiency' is a very complex conception.

"If I had time I should like to try to think further about the scheme, and lend what aid I could to it; but unfortunately I am so fully occupied by trying to finish a book for the sake of which I retired from the Embassy at Washington that I am, to my regret, unable to coöperate with you. I shall be grateful for any information as to the results you can send me if you have leisure."

"P. S. [dated two weeks after the letter].

"I have kept the papers some weeks in order to see if I could help you. I have dealt with the states and provinces of the United States and Canada, though roughly. But I have found the difficulty of any adequate arrangement of the extra-European world in point of efficiency impossible in respect of the extraordinary mixture of races. Turkey and India are nuts too hard to crack; . . . In South America it would be easier, because the races are really so mixed that a new race results. But in India they live side by side and are quite different. [Here follows a classification of South America.] It is quite interesting and makes one think. But after all it is the natural race divisions rather than territorial divisions that count everywhere outside western Europe. Even in Russia and the Balkans, race doesn't correspond to territory."

These various letters give a good idea of the general opinion of the attempted classification. As an illustration of the difficulties to be overcome, let me quote from an Italian contributor, an anthropologist: "What is the standard of these various civilizations? Yours, as it seems to me, is an European one, and this, I think, is a very limited one and cannot solve the problems." A similar objection is expressed in a letter from an English anthropologist. Quoting from my letter, he says:

" 'The power of self-control, high standards of honesty and morality, . . . high ideals, respect for law' are eminently characteristic of many savage and barbarian peoples, notably the North American Indians, and in my opinion these latter stand very much higher than the average American citizen, but the latter lead in 'the power to lead and control other races, and capacity for disseminating ideas . . . inventiveness, highly developed systems, etc.' "

I recognize the force of these comments. Races certainly differ greatly, even though they happen to dwell in the same physical environment. Moreover, people of a given race who live under the same environment may differ widely because of diversity in religion, government, or institutions. Furthermore, the definition of civilization here presented does not pretend to be perfect. It is a European definition. Yet it is also a world-wide definition. The contemplative Hindu may perhaps be a higher type than the aggressive citizen of western Europe, but the contemplative type has made relatively little impression upon the world as a whole. If we turn to antiquity, the people who have left their impress are those who had this so-called European activity. The Greeks and Romans had it to a marked degree. The people of India had something of it when they wrote the *Vedas*. Gautama had it when he founded the Buddhist religion. He was contemplative, but yet he had the qualities expressed in our definition. He was preëminently possessed of "high ideals, respect for law, inventiveness [in the broad, non-technical sense], ability to develop philosophical systems, and the capacity to dominate the less civilized parts of the world." He dominated through ideas, not force. The Jews had this same power. Such men as St. Paul, although not aggressive in the ruder European sense, were unsurpassed in "the power of initiative, the capacity for formulating new ideas and for carrying them into effect, the power of self-control, high standards of honesty and morality, the power to lead and control," and "the capacity for disseminating ideas." The reason for regarding the standard here set forth as European is that Europe is today its great exponent. In the past, however, not only did Rome, Greece, Palestine, and even northern India possess it, but Egypt, Mesopotamia, and Carthage all displayed it. China, too, in her days of early greatness, and the wonderful Maya people of Yucatan, the only ones to develop the art of writing in America, were animated by the same active, stirring, "European" qualities. That is why we remember them, but have forgotten most of their contemporaries.

Granting that our definition of civilization is imperfect, but admitting that it includes the qualities which are of greatest importance in causing a nation to impress itself upon the world, let us now proceed to ascertain how these qualities seem to be distributed. The fifty contributors whose classifications could be used have been divided into five divisions as follows: (1) 25 Americans, 1 of whom is a Canadian; (2) 7 British, 1 of whom is an Australian; (3) 6 Germanic Europeans, 4 of whom are Germans, 1 a Swede, and 1 a Norwegian; (4) 6 Latin Europeans, namely, 2 Frenchmen, 2 Italians, 1 Spaniard, and 1 Portuguese, with whom has been included 1 Russian because there is no other group in which he fits more appropriately; and (5) 5 Asiatics, 2 of whom are Chinese and 3 Japanese. A third Chinese contribution was most unfortunately lost in the mail. The average opinion of each of these five groups is given in the tables in the Appendix. To obtain the final rank of each country the averages for the five groups have again been averaged. Thus each race, or at least each of our five divisions, has equal weight in determining the figures on which will be based the map used in later discussions. The opinion of twenty-five Americans, for example, has no more weight than that of five Asiatics. This may seem unfair, but on the whole it seems to be the method best calculated to give a reliable result. All of us are inevitably prejudiced. The Americans put America, especially its more backward parts, higher than is correct. The Asiatics put their own countries too high. By giving America and Asia equal weight and by dividing Europeans into three groups animated by different ideals and different sympathies, we are able largely to eliminate the effect of racial prejudice. The final results are summed up in Figures 23 to 28, but I shall defer comment upon these for the present.

In the tables in the Appendix the countries of Europe, North America, and Asia have been divided into groups corresponding as nearly as possible to race, while those of Australia, Africa,

and South America have each been put in a single group because
the racial differences are either not strongly marked, as in
Australia and South America, or are highly complicated as in
Africa. Under each group the country whose final rank is highest
has been placed first and the rest in consecutive order. To begin

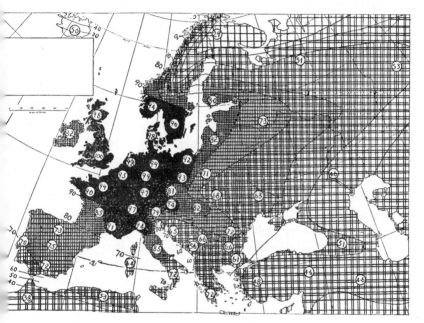

Figure 23.   The Distribution of Civilization in Europe

with Europe, England heads the list. It is the only region placed
in the first rank by every contributor. Northwestern Germany,
which includes Berlin, comes next. The Germanic Europeans
and the Asiatics all place it in the first rank. The British almost
invariably do so, and their opinion, 9.9, averages the same as
that of all divisions. Among the Americans and Latins several
place this part of Germany in the ninth class instead of the
tenth. Hence, its average position according to them is about

9.8. Such slight differences have little significance, for 98 per cent is almost as good as 100 per cent. Yet they suggest that in 1913 people who live in other parts of the world were on the whole not quite so sure of the Germans as of the English.

Following northwestern Germany we find central Germany, then Scotland, Denmark, Holland, and so on to northeastern Germany. Here for the first time we come to a region which is placed by one group, the Latins, in a rank lower than 9.0. Yet even here the Latins do not assign a rank lower than 8.5—and everything above this ranks as "very high." Nowhere in these first fourteen regions does the greatest diversity amount to more than one degree on our scale of ten, or 10 per cent on the scale of 100. This is comparatively insignificant, for it means that while the difference between the highest and the lowest may be 10 per cent, each of them usually differs from the average by only about 5 per cent. Thus so far as numbers 1 to 14 are concerned the difference of opinion among Americans, British, Germans, Latins, and Asiatics is almost negligible. All alike rank these regions very high.

Coming to Ireland, a country which, for lack of any other suitable group, is placed with the Teutonic regions, we find much more diversity of opinion. The English, presumably because it is part of their own empire, and the Asiatics, perhaps because it seems to them like a part of England, place it very high, with a rank close to 9.0. The Germans, on the other hand, place it at 7.0, scarcely above the medium grade, while the Americans and Latins place it near 8.0, which means that compared with the world as a whole they think that Ireland stands high, but not very high. The fact that all the Teutonic regions except Ireland and the Austrian Alps rank above 9.0, and that these two, which are partly Teutonic, stand in the high group near 8.0, suggests that race is a dominant factor in determining the status of civilization. The same suggestion is enforced when we note that among the Romance nations the most Teutonized portions stand highest. Yet the fact that innate racial differences may be of great and even overwhelming importance, as I have shown in *The Character of Races*, by no means alters the fact that geographical conditions also play a highly important part.

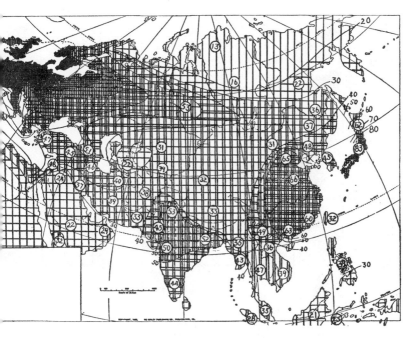

Figure 24. The Distribution of Civilization in Asia

The Romance nations of Europe seem to possess much more diversity of civilization than do the Teutons. Even if Albania and Montenegro be omitted as not being truly Romance, the range is from medium in Corsica and Sardinia to high in many regions, and very high in parts of France and northern Italy. Some places, such as southeastern Spain and southern Italy, are rated exceptionally high by the Asiatics, although the other contributors agree quite closely.

Among the Slavic nations central Russia stands at only 6.2 in American opinion, possibly because Jewish atrocities were freshly in mind when the classification was made. In general there can be no question that the Baltic Provinces and Bohemia stand at the top, while southeastern and northeastern Russia are at

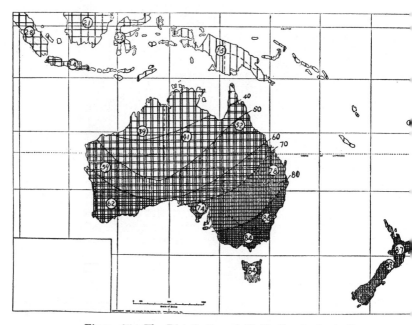

Figure 25. The Distribution of Civilization in Australia

the bottom. The European region whose rank is most doubtful is southern Finland. The Asiatics reduce it to about 6.6, while the Germans rate it at 9.4. Even in so extreme a case the average opinion is not open to much criticism. The final rank is about 8.0, which puts Finland at nearly the same level as Bohemia, the Baltic Provinces of Russia, Ireland, and the Austrian Alps.

It would be interesting to go through all the tables and point out their special features, but this must be left for the reader to do by himself. Only one or two additional points can be indicated. In Asia the first thing that strikes one is the great diversity of opinion as to Japan and China. This is due to the

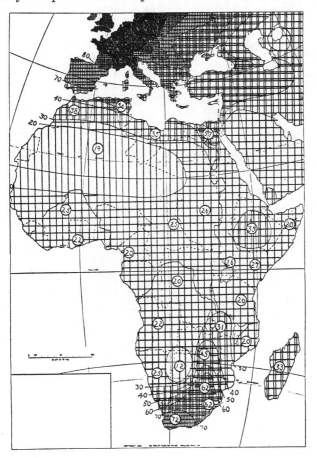

Figure 26. The Distribution of Civilization in Africa

fact that the Japanese and Chinese place their own countries much higher than do the people of other races. This is natural. The surprising thing is rather that these people, with their

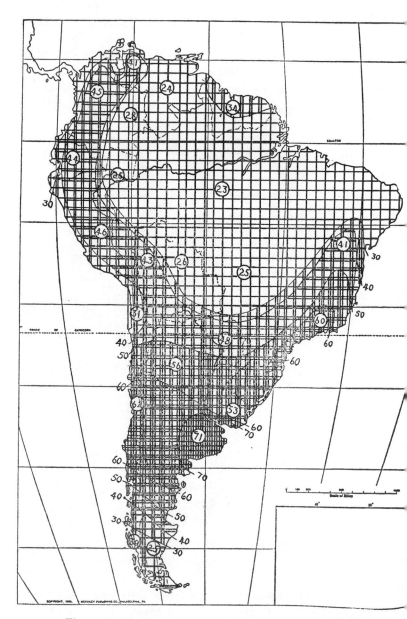

Figure 27. The Distribution of Civilization in South America

justifiable pride in a great past, do not place their own countries at the very top. They recognize that Europe and North America have in certain ways surpassed them. Aside from Japan and China the agreement of the different groups as to the position of Asiatic countries is on the whole surprising. Only in rare cases does the opinion of any one group depart from the average opinion by more than a single degree on the scale of ten. Two of the few exceptions are Rajputana and the Syrian Desert, both of which are placed exceptionally low by the Asiatics.

In Australia and North America the diversity of opinion is much greater than in the older continents. It reaches a maximum in Canada, especially in Alberta, and in southern Greenland, and Iceland. In such places opinion ceases to be of any special value. People simply do not know what sort of conditions actually prevail. In Australia, likewise, the comparative newness of the country causes some people to overrate it and others to depreciate it. It has not yet reached the stable equilibrium which gives the world as a whole a well-defined opinion. In the United States there is no more diversity of opinion than in Europe, for that country has taken its true place among the nations. In Australia and Canada, on the other hand, outsiders who live far away and who are not racially connected with the inhabitants are still sceptical as to whether those countries are actually destined to rise as high as their people claim.

Thus far we have drawn attention to differences of opinion more than to agreements. Let us now examine the matter in another way, which brings out the essential agreements. We will take North America because this is the continent as to which there is the greatest diversity of opinion. Figures 28 to 33 present a series of maps based on the average opinion of all divisions of contributors and on the individual opinion of each division respectively. Here, as in Figures 23 to 27, the rank of each region has been written in its proper place on the map. Then lines have been drawn in such a way as to separate the areas

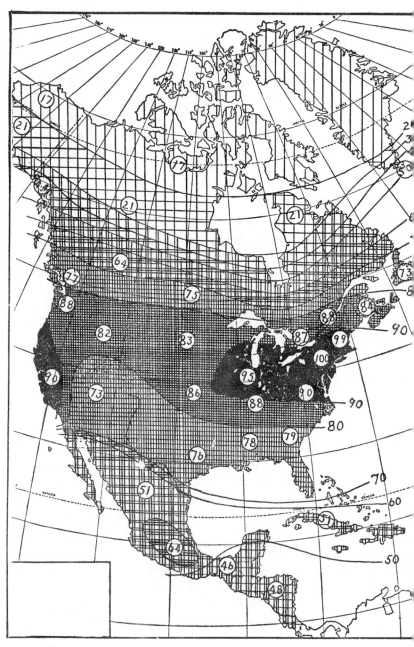

Figure 28.   Civilization in North America, According to all Contributors

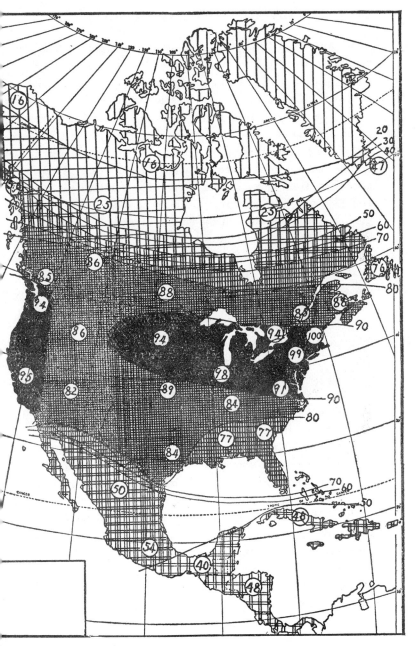

Figure 29. Civilization in North America, According to Twenty-five Americans

above and below 9, above and below 8, and so forth. The areas ranking above 9.0 have been heavily shaded, those from 8.0 to 9.0 less heavily, and so on until places below 2.0 are almost unshaded. This gives a map of civilization whose main features can readily be grasped. In Figure 28, which represents the average opinion of all groups of contributors, certain main features stand out prominently. They are (1) an area where civilization declines rapidly northward in British America; (2) a high area extending from east to west across the northern half of the United States and a narrow southern band of Canada; (3) an exceptionally high area in the northeastern United States from the Mississippi to the Atlantic Ocean; (4) a similar small, high area on the Pacific coast; (5) a bight of lower, but not of low, territory extending into Nevada; (6) a rather rapid decline south of the United States, which is interrupted by (7) an area of slight improvement in central Mexico.

Let us now turn to Figure 29 and see how the opinion of Americans differs from that of the world as a whole. The general aspect of the two maps is the same, for every one of the seven features just mentioned can be detected. The older states along the Atlantic from New England to the northern Gulf States have almost identically the same rank in both maps. The chief difference lies in this: the Americans have a higher idea of the new parts of their country and of the new parts of Canada than have the people of other places. They have a correspondingly low idea of the countries immediately to the south of them. In Central America their opinion does not differ greatly from that of the rest of the world, but they place the West Indies and Mexico relatively low. In the West Indies this is probably due in large measure to the fact that in order to prevent internal convulsion the United States has repeatedly intervened in Cuba during the brief period of self-government enjoyed by that country. The low estimate of Mexico probably arises from recent revolutions which at the time when the classification was made were producing a most unpleasant impression in the United States. In Europe and Asia the disorders of Cuba and Mexico attract relatively little attention, and the contributors to our classification probably thought of these countries as they were during their periods of long peace.

Now let us study British opinion as expressed in Figure 30. One of the most noticeable things is the English opinion of the southern United States and especially of the southwestern deserts. No patriotic feelings, either conscious or unconscious, lead them to believe that high conditions extend to the Mexican boundary beyond which Americans believe that there is a great and sudden change. Their opinion of the Atlantic portions of the United States is almost identical with that of the Americans and of the rest of the world, and they agree with other races in their opinion of the regions south of the United States. They believe, as does almost everyone, that the Pacific coast of the United States stands higher than the western part of the interior, but they do not think quite so well of it as do the Americans, although the difference is slight. The places where their national pride comes into prominence are in Newfoundland and the Canadian Northwest. It is interesting to notice that while they place Newfoundland and British Columbia higher than do the Americans, they do not have so high an opinion of the Maritime Provinces nor of Alberta, Saskatchewan, and Winnipeg. In the latter regions this is easily explicable. A great number of Americans have gone to Winnipeg, Alberta, and Saskatchewan during recent years, and their glowing reports have made upon Americans a great impression which has not yet reached England so strongly, and has scarcely touched the rest of the world. Americans perhaps think more highly of the Maritime Provinces than do the English because many of the most energetic Nova Scotians and other provincials come to the States and display marked ability. To cite a case which has come under my own observation, the students from Acadia College in Nova Scotia,

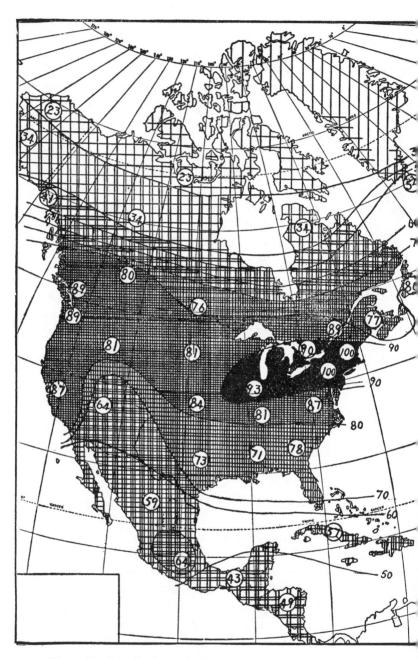

Figure 30. Distribution of Civilization in North America, According to
Seven British Contributors

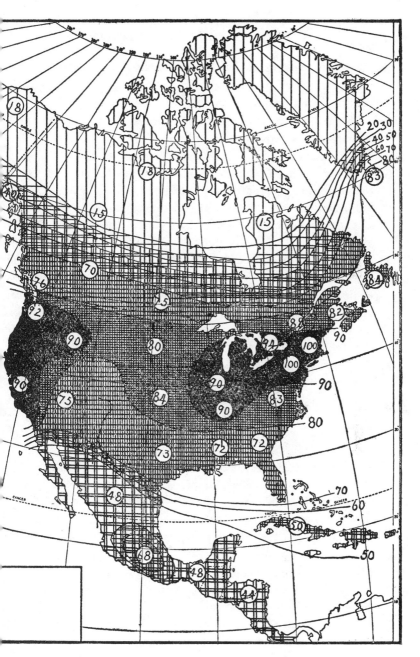

Figure 31. Civilization in North America, According to Six Germanic Europeans

who carry on work in the Graduate School of Yale University, for some years maintained a higher average grade in their studies than did those from any other institution.

In the German map, Figure 31, the seven general features already enumerated can readily be detected, just as in the American and English maps. The differences are largely matters of detail. Such a feature as the assignment of a rank of 9.0 to the northern Rocky Mountain states of Idaho, Wyoming, and Montana, and of only 8.0 to the Dakotas, Minnesota, Nebraska, and Iowa is probably due to lack of familiarity with the interior of the continent. Why Newfoundland should stand so high is not evident, unless it be because the name has long been known and is very familiar. In the high position of Iceland racial pride is again evident, for the Icelanders are close akin to the Germans.

The Romance peoples present a map (Figure 32) which again shows all the main features. They know, however, even less of the recent development of the western United States and of the Canadian Northwest than do the Germans. To them Alberta and Saskatchewan are apparently undeveloped tracts where trappers and Indians still roam, for otherwise they surely would not assign so low a rank as 4.4. Newfoundland for some reason is also placed low, only 5.2, and Greenland still lower, 3.4. On the other hand, just as the presence of Germanic people in Iceland and Greenland draws the high lines far to the northeast in the Germanic map, and as pride of country bulges the high lines to include the whole of the United States in the American map, so a similar racial pride causes Cuba and the Creole region of Louisiana to be much higher on the Latin map than on any other.

Last come the Asiatics with a map, Figure 33, which is surprisingly like the other maps. They do not yet realize what has happened in the Canadian Northwest, but their opinion of eastern Canada and Newfoundland is close to the average. The place where they differ chiefly from others is in their relative ideas of the northern Rocky Mountain states to which they assign a rank of only 7.6, and of the southern states which they place all the way from 8.2 to 9.4. Apparently the eastern United States seems to them highly progressive in all parts, and they do not attach much importance to the presence of the negroes, a disadvantage which Americans themselves feel strongly.

Taking these maps as a whole we see that they all give the same general impression. Where one group goes to a great extreme, as in Newfoundland or Greenland, some other is likely to go to the opposite extreme. Thus the eccentricities of judgment displayed by one race are largely counteracted by those of another. When all are averaged, the number of inconsistencies is greatly diminished. To Americans and English it may seem that Alberta and Saskatchewan, for example, should have a rank higher than is here given them, but we are forced to admit that final judgment is not possible until these regions have been well populated for a generation or two. Taken as a whole Figure 28 seems to contain no important inconsistencies, even though one might wish to change the rank of certain places in which he is particularly interested. The map represents the judgment of thoughtful people in many countries. If fifty other equally well-informed people representing an equally large number of countries in three separate continents were chosen to make another classification, it is highly improbable that their map would differ from Figure 28 so much as this map differs from the others. The chances are that the two would be so nearly alike as to be indistinguishable. Inasmuch as there is more diversity of opinion in regard to North America than to any other continent, the maps of the others approach still more closely to a true representation of the opinion of thoughtful people all over the world. The maps are by no means perfect, for human judgment is fallible. Yet they at least present so close an approximation to the truth that we may use their general features without danger of error.

During the World War and the succeeding period of hot

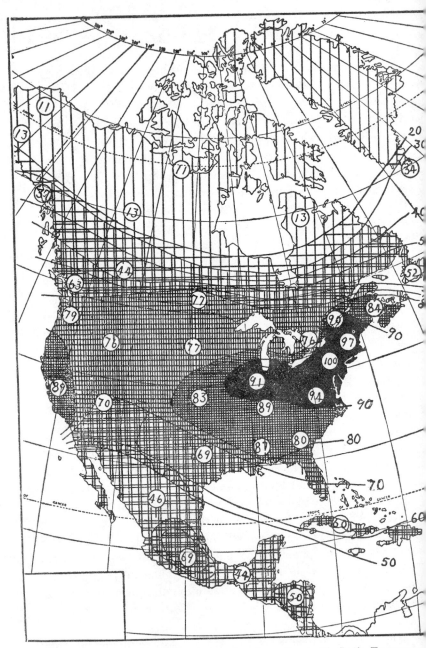

Figure 32. Civilization in North America, According to Six Latin Europeans and One Russian

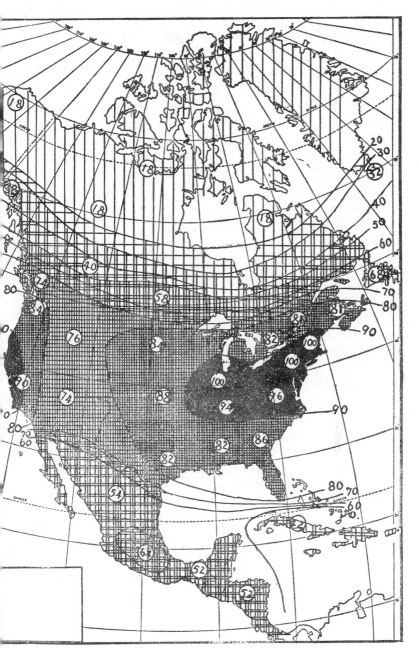

Figure 33. Civilization in North America, According to Five Asiatics

prejudice it seemed to many people that the truth of this last statement was doubtful. But soberer thought today shows that this is by no means the case. During the war the steadfastness and ability with which the various races fought was almost directly in harmony with their position in our map of civilization. The Portuguese, with a rank of 70, had to be taken out of the main line of battle, for they could not withstand Germans, who rank above 95. The Austrians, Poles, Italians, and Russians, not to mention the Asiatics and Africans came out of the war with less military prestige than the British, French, and Germans. Moreover, since the war the European countries shaded black in Figure 23, that is, those ranking highest in civilization, have recovered more rapidly than their neighbors of lower rank. France and Belgium in spite of being ravished, and Germany, in spite of her financial debacle, have strengthened their positions in many respects, and are pushing ahead with remarkable energy. Russia, on the other hand, remains in a curious state of suspended animation, perplexing the world by her combination of inertia and radicalism. The Soviet system is perhaps only nominally at the root of her troubles: the wholesale exodus and slaughter of the old leading classes have probably done irreparable damage whose full results may not appear for many years. But back of both these facts lies something in Russian character, something which before the war held that country down to a rank of from 46 to 73 in Figure 23. Spain, Italy, Hungary, Greece, and the rest of southern and eastern Europe have likewise behaved as one would expect of countries on the level of progress indicated by their respective positions on our maps of civilization. Thus, on the whole, the World War and its aftermath seem to confirm the value of our method of estimating human progress.

CHAPTER XII

VITALITY AND EDUCATION IN THE
UNITED STATES

WE now have before us maps of the world depicting the distribution of climatic energy and of civilization. Our ultimate object is to compare them, and see whether they are causally related. It will add to the certainty of our results, however, if we first apply an independent, statistical test to each of our two maps. Since the methods of compiling statistics vary greatly, our results will be most reliable if we confine ourselves to a single country. The United States is easily the best for this purpose. It possesses a large and highly varied area which tends to produce diversity. Its census and such organizations as life insurance companies furnish data compiled according to the same method in all parts. It is homogeneous in government and institutions; equal opportunities are open to all; and the same ideals and methods prevail almost everywhere. Moreover, no part has been devastated by war or any other great disturbance for nearly two generations; the people have moved freely from place to place; and there has been a constant tendency to foster a single type of culture. From all these points of view no other equal area is so nearly uniform. Racially the country is of course complex, but until the last few decades the great majority of the people have been Nordics from northwestern Europe. The various types have mixed to such a degree and have become so strongly Americanized that the native white population is everywhere similar. The only large elements which tend strongly toward diversity are the negroes and the recent immigrants from

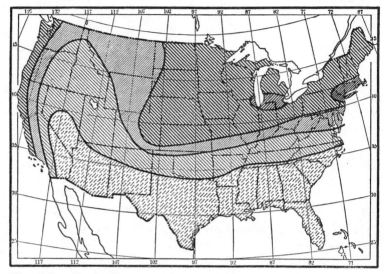

Figure 34. Climatic Energy in the United States on the Basis of Factory Work

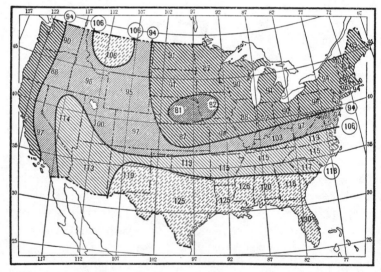

Figure 35. Vitality in the United States, According to Life Insurance Statistics

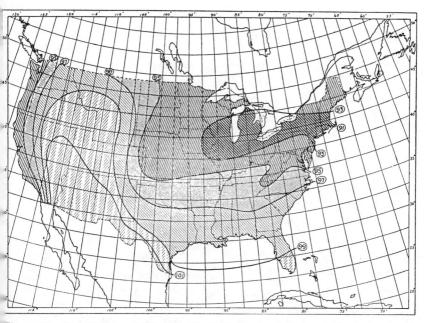

Figure 36. Distribution of Climatic Health on Basis of Seasonal Variations in 33 Cities, 1900-1915

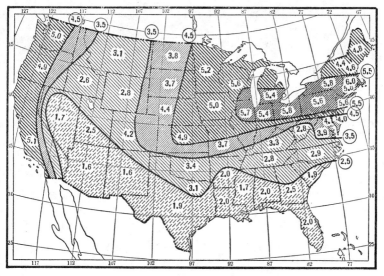

Figure 37. Civilization in the United States. The Numbers Indicate the Relative Rank on a Scale of 6.0 According to 23 Contributors. The Highest Possible Rank is 6.0 and the Lowest 1.0. This Scale is Entirely Independent of the One Used for the World as a Whole

non-Teutonic countries. If these are eliminated, as they are to a large degree in the maps that we shall consider, essentially the same degree of energy and civilization ought to prevail throughout the country except where geographic surroundings, the presence of some special institution, or some other disturbing factor makes itself felt.

One of the best tests of energy is vitality. We have already seen that the death rate varies from month to month in close harmony with variations in the strength of factory operatives. Let us now see whether there is similar geographic harmony. Figure 34 shows the distribution of climatic energy in the United States. It is a revision of Figure 22, but on a larger scale. Figure 36 is a new map of climatic energy based on seasonal variations in the death rate as described in Chapters VII to IX. The general similarity of these two maps compiled in absolutely different ways is one of the strong reasons for believing that they give a true idea of the distribution of climatic energy. Where either one is heavily shaded we should expect people to be strong, vigorous, and long-lived. We might test this expectation by ordinary mortality data, but such data make no allowance for age nor for such conditions as the preponderance of unhealthful cities and manufacturing in the Northeast, and the natural selection of more vigorous types as migrants to the West. A more reliable test is furnished by insurance companies, for their policy holders are much more homogeneous than the population as a whole, and full allowance is made for age. A comparison between the actual deaths among policy holders and the deaths expected on the basis of standard life tables gives a good idea of the vitality of a community. Figure 35 shows such a comparison. It is based on the combined experience of three prominent life insurance companies whose officials have kindly placed in my hands the figures for hundreds of thousands of people arranged by states.

On the map the figures indicate the actual number of deaths compared with the average for the whole country, which is taken as 100. The country has been divided into five grades corresponding to those of the energy map. The two heaviest shadings include regions where the mortality is distinctly less than would be expected from the actuarial tables. The third degree of shading indicates conditions a little better than the average; the fourth, high mortality; and the lightest, very high. The individual maps for each of the three companies are closely similar, which indicates that the general features are not due to any special policy of one company. Indeed, so far as the policy of the companies is concerned, the southern states ought to present a better record than the northern, for the restrictions upon the issuance of policies are there more rigid. For instance, one company entirely refuses to issue any policies in certain unhealthful sections of the South. Elsewhere only people engaged in special kinds of healthful occupations are accepted. Moreover, in the places where risks are accepted upon the same terms as in the North, the medical examination is often more strict, especially in respect to preventable diseases such as tuberculosis.

Various other considerations differentiate one region from another. For instance, in cities, especially in manufacturing cities, the death rate is higher than in country districts, and this causes New York State to be unduly high. Miners are an especially precarious class from the standpoint of the insurance companies, and, therefore, are excluded by at least one of our three companies in some of the more remote western states, and are accepted only in small numbers by the others. They may account for the poor position of Montana, but the neighboring mining state of Idaho ought in that case to be equally bad. In all the more important respects the tendency would be to cause the death rate in the South to be lower than in the North, were it not for the disturbing element of physical weakness due to climate. In tropical countries the figures are far worse than in the southern states, which is what would naturally be expected. The fact that sick people often go west for their health does not enter into the problem. Such people cannot obtain insurance. If they have been insured before they become sick, they are reckoned as belonging to the place where they lived when the policies were issued, and not to the place where they die.

It would be interesting to enter into further details, but space does not permit. The outstanding feature of the vitality map is its agreement with the maps of climatic energy. As people die in greatest numbers during months when their energy is low, so they die in parts of the country where their energy would be expected to be diminished on account of the climate. In Figures 34, 35, and 36 conditions are best in the North from New England to Kansas. Westward and southward they become less favorable. The decline in health is probably due in part to the direct physiological effects of climate, and in part to its indirect effects upon sanitation and other methods for the promotion of health. The direct and indirect effects of climate almost inevitably go together, for where people's energy is great, they are quick to adopt new means for the prevention of disease and the improvement of health. Moreover, in dry, and still more in warm, regions it is easy to tolerate unsanitary methods of disposing of sewage, which thus pollutes the water supply. Yet all these things would apparently lose part of their importance were it not for the weakening effect which certain climatic conditions unquestionably produce, as is so well proved by the variation of the death rate from month to month even in places where the conditions of health are most carefully looked after. Here we must leave the matter. The general agreement of the vitality map with the map of climatic energy affords strong evidence that the climatic map is correct.

We are now ready to test our map of civilization in the same way that we have tested the map of climatic energy, that is, by comparing it with a map based on statistics. Strength and energy of character cannot easily be reduced to statistics, for most of the conditions and activities for which we possess exact data depend too much upon outside circumstances and not sufficiently upon the actual qualities of the people. Prof. Mark Jefferson has studied the matter carefully, and I have borrowed freely both from his published and unpublished work. It might seem as if such things as railroads, the number of letters, the amount of manufactures, or other similar matters might furnish a good clue to the intellectual capacity and cultural development of a people, but unfortunately this is not so. Take the case of railroads. Nevada has more miles of railroad in proportion both to the number of inhabitants and the inhabited area than any other state in the Union. In 1920 it had 198 miles for every 10,000 inhabitants, while Rhode Island had only 3.3, and Massachusetts 5.5. This does not mean that Nevada is more progressive than New England. The case is like that of a desert through which a cowboy was riding when he met a friend.

"What you doing here?" asked the friend.

"Nothing," was the answer. "I'm just crossing this here desert because it's here."

In the same way the railroads cross Nevada because it happens to lie between the East and the West. If it were uninhabited, or peopled by savages, it would still have many railroads.

The letters sent out by a community furnish a criterion of its state of civilization, but even this must be used with much caution. A hundred letters sent by a Chicago mail-order house in response to orders averaging two dollars apiece are no more significant than a single letter from Detroit in answer to an order for an automobile. In this case, as in many others, a concentration of activity in certain regions may occur without any correspondingly high ability or culture. The same is true of manufactures. Doubtless, manufactures generally develop wherever a community rises to a high state of civilization, and the manufacturing processes and all that goes with them are in turn a help in the development of a still higher civilization. Nevertheless, the accidents of position, or the presence of natural resources are commonly supposed to cause equally progressive and competent communities to differ enormously in the number of factories. We shall examine this supposition later.

We are forced, therefore, to turn to something more personal. Illiteracy and education are fairly good tests, for they depend largely on the immediate surroundings of each individual. Illiteracy would answer the purpose excellently were it not that people have moved about so much in recent years. Education, however, is still a local matter, especially in a country like the United States. Each state, and often each county or town, decides for itself how much it will spend for schools, how long they shall be open, and how stringently attendance shall be enforced. Hence, the schools form an unusually delicate test of the real character of a community. The quality of an educational system cannot, indeed, be measured exactly by statistics. Nevertheless it is fairly well indicated by certain conditions for which accurate statistics are available. For this purpose I have selected the following data for the year 1920 from the reports of the United States Commissioner of Education and from the census.

1. Percentage of young people seven to fourteen years of age who attend school. Colored children, those born in foreign countries, and the children of foreign-born parents are excluded in order to make our data as homogeneous as possible. The percentages range from 82.5 in Utah and 79.6 in Idaho to 66.9 in Maryland and Georgia and 65.4 in Louisiana.

2. Percentage of young people eighteen years of age who graduate from a four-year course in either a public or private high school. Since colored and foreign-born persons are not distinguished from native whites in the data for high school graduates I have assumed that all the graduates were whites, and have calculated the percentage of high school graduates accordingly. This of course gives the South a certain advantage and thus compensates for the fact that in using the next two

criteria it has been necessary to include colored pupils in one case and colored teachers in the other. It must be remembered, however, that the foreign-born population of the Northeast tends to lower the standards there just as do the negroes in the South, although not so much. The maximum percentage of high school graduates is reached in Maine 28.6, New Hampshire 27.8, and Oregon 26.2. The minimum is found in Georgia 3.9 and South Carolina 2.2, but when the allowance described above is made for colored people the figures become 6.6 and 4.5.

3. Average number of days per year that each pupil in the public schools was actually in attendance. Good schools as a rule are in session fairly long, and always insist on a high degree of regularity in attendance. The range here is from 151.3 in New Jersey and 149.6 in Massachusetts, to 76.0 in South Carolina and 76.9 in Mississippi.

4. Average salary of teachers. Unless teachers are well paid, the more capable young people are likely to give up teaching for some more lucrative and less nerve-racking occupation. Contrary to general belief, teachers are paid at about the same rate in the Northeast and the far West, the maximum averages being $1282 in New Jersey and $1262 in Massachusetts compared with $1375 in Oregon and $1281 in Washington. In the intermediate regions the general scale of recompense falls lower, dropping to $1081 in Illinois, which outranks all its neighbors except Ohio, and to $696 in South Dakota. In the South the level is very low with a minimum of $426 in Georgia and only $291 in Mississippi.

5. The excess of young men over young women among students eighteen to twenty years of age. This last item may at first sight seem unimportant, but its significance grows as one studies it. In regions where the standards are low there is a tendency for the older boys to drop out of school, while the girls keep on in order to become stenographers, bookkeepers, or, especially, teachers. Maryland and Utah have the greatest excess of

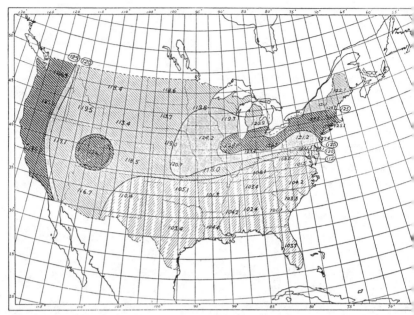

Figure 38. Distribution of Education in the United States in 1920 on Basis of F:
Factors Equally Weighted

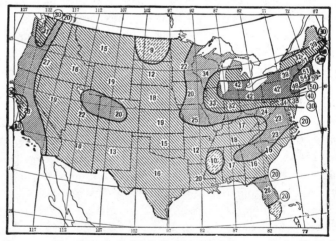

Figure 39. Percentage of Gainfully Employed Persons in the United
States Engaged in Manufacturing, 1919

boys over girls, and Wyoming, Montana, and the Dakotas the greatest excess of girls.

In using these five criteria each one was given exactly the same weight, that is, the census figures for any given criterion were all lowered or raised proportionally, so that for each criterion the difference between the lowest and highest figures was the same. As the number of criteria was increased it was noticeable that the resultant map, Figure 38, became more regular. In its final form the map shows that in a relatively narrow belt extending from southern New England through New York, but not Pennsylvania, and continuing to Illinois, the general conditions of education are excellent. The same is true in Utah and on the entire Pacific coast. In those places, on the whole, a large percentage of the children are in school, a large proportion graduate from the high school, the school year is long and the pupils are regular in attendance, the teachers are well paid, and many boys as well as girls continue to study after the age of eighteen. One or another of these conditions may break down in any individual state, but when all are taken together the heavily shaded areas of Figure 38 rank high.

One of the most noteworthy features of Figure 38 is the relatively low position of the Dakotas and the other states in the western part of the Great Plains, and also the similar position of all the Rocky Mountain states with the exception of Utah. The comparatively recent settlement of some of these states explains the situation in part, but Washington is almost as new as the Dakotas or New Mexico and yet stands in the first rank. The sparsity of population is another factor which hinders education. Utah, however, has a population decidedly more sparse than that of the Dakotas and Wyoming, but ranks 126 where they rank from 113 to 119. Its density of population is scarcely greater than that of New Mexico, whose rank is 110. The relatively poor conditions in the Dakotas and Wyoming find a partial explanation in the fact that the population is more completely agricultural than in almost any other part of the United States. There are no large cities and even villages are comparatively scarce or small. The great majority of the people live scattered over the vast plain, each family on its own quarter section. Hence, the children are obliged to go long distances to school. Muddy roads often make this difficult in the spring, while the severe storms of winter are an even greater obstacle. In spite of this, the Dakotas stand at the top in the literacy of their people, so that the school system must be fairly efficient.

In New Mexico quite the contrary conditions prevail. A part of the people, to be sure, live on widely scattered ranches where the children cannot go to school. By far the larger number, however, live in compact settlements where the houses are grouped in a comparatively small area because of the necessity for using a common water supply for irrigation, or else because of mining industries. Under such conditions, schools can be maintained more easily than on huge, townless plains like those of the Dakotas. The same conditions prevail in Utah and Nevada, and to a less extent in Idaho, Wyoming, and Montana. Nevada's rank of 117 may be less creditable than South Dakota's of 114. Yet the fact remains that the Dakotas are lower than would be expected, while Utah is surprisingly high.

The proud position of Utah is presumably the result of Mormonism. The leaders of that faith have had the wisdom to insist on a thorough system of schools, and have obliged the children to attend them. The "Gentiles" have in self-defense been forced to do equally well, and the result has been admirable. Whatever one may think of Mormonism as a religious belief, it must be credited with having accomplished a remarkable work in spreading a moderate degree of education almost universally among the people of Utah. Without its influence, the rank of Utah would probably be about 118, that is, between Colorado (119), on one side, and Nevada (117), on the other. I emphasize this because it shows how clearly our maps reflect the influence of any peculiar condition. Manifestly, the distribution of education throughout most of the United States does not depend upon the influence of any particular institution, for essentially the same institutions prevail everywhere. Yet in the map of education, Utah is conspicuous because it is strongly influenced by a unique American institution which is limited to one small area.

Another factor which would be expected to bear an important part in determining the distribution of education is the presence of the negro. Doubtless this has a pronounced effect, for an inferior race inevitably retards a higher. Yet the map indicates that other factors are equally important. Not only is education at a low ebb in New Mexico which has few negroes, but also in West Virginia which is comparatively free from both negroes and Mexicans. Moreover, Texas, where the colored people form only a fifth of the population, stands lower than South Carolina and Mississippi, where half the people are colored. Another significant fact is that the education of negroes varies from state to state almost as does that of whites. Although a smaller proportion of negroes than of whites go to school, the two races are well or poorly educated in the same places. In Figure 38 the figures for states with a moderate number of negroes, such as Virginia with 30 per cent, give a fairly correct impression, but the impression is not so accurate in those states where there are many negroes. The reason is that our index figures for states with many negroes are too high by reason of our assumption that all the high school graduates are white. Disregarding these minor discrepancies it seems quite clear that the degree of education is not proportional to the number of colored people. The southern states differ among themselves because of special circumstances such as good or bad laws, but all stand low because of more general factors. These factors give to Figure 38 its general character, and seem to make it as good an epitome of the general distribution of culture in the United States as we are yet able to obtain on the basis of the statistics of a single activity. Nevertheless, it can scarcely be doubted that the general level of the South is much lower than it would be if there never had been any colored people there, but that would not change the general aspect of Figure 38.

Let us now compare our educational map with Figure 37, which represents the distribution of civilization in the United States according to the opinion of 23 people. All of these except Ambassador Bryce were Americans. They grouped the states and provinces of the United States and Canada into six classes, number 6 being the highest. Massachusetts is the only state invariably placed in the highest class. New Mexico and Arizona, which stand lowest, have an average rank of 1.6. In order to judge how much reliance to place on the classifications, I took the first ten that were received and averaged them, and when ten more had come to hand averaged them also. Somewhat to my surprise, and much to my pleasure, the two sets of averages were practically identical. How much they differed may be seen in the Appendix. The average difference is only 0.2, and the maximum 0.6. The agreement of the two sets probably indicates that any other group of equally well-informed persons would have made essentially the same classification. To be sure, in spite of several attempts, I was unable to obtain any contributor in the states west of Minnesota or south of the Ohio River. Local prejudices, however, have probably not exerted much effect on the final results, for California stands in the highest class, with practically the same grade as Minnesota and Iowa.

A comparison of the maps of civilization (Figure 37) and education (Figure 38) is interesting. In general aspect the two are similar. Both have two high areas, one in the northeast and center, and the other on the Pacific coast. In both there is a decline from north to south. Another common feature is a tongue of high conditions jutting out toward Kansas. This eastern high area and its western counterpart are almost identical with those of the map of civilization. In both maps Massachusetts takes first place. In the map of progress Connecticut, New York, and Ohio come next, all three being equal: in the educational map the rank is Connecticut, New Jersey, Washington, California, Ohio, Utah, Oregon, and then New York. Judging by this our eastern contributors to the map of progress did not quite do justice to the Pacific coast. Nevertheless, the differences among the states just mentioned are too slight to be important.

The only real discrepancies between Figures 37 and 38 are the high position of Utah in the educational map, which has already been explained, and the absence of the low tongue jutting into Nevada. This is perhaps due in part to an undue lowering of the general educational level of the Southeast by reason of the negroes. Also the contributors to the map of progress probably paid too much attention to aridity, sparseness of population, and the ephemeral character of mining towns, and not enough to the fact that the inherent quality of the people of newly settled regions is almost invariably high. The process of selection incident to migration, as I have shown in *The Character of Races*, goes far to insure such quality, provided the migration is unassisted and is beset with sufficient difficulty.

In concluding this chapter, let us look at another criterion which is often supposed to be closely associated with civilization. This is the percentage of the population engaged in manufacturing. It is often asserted that manufacturing is most predominant in places where supplies of coal and iron are present, but this is a mistake. A mathematical analysis of the states of the United States by means of correlation coefficients shows that there is no relation whatever between the amount of coal mined per inhabitant and the percentage of the population engaged in manufacturing. In the world as a whole the same is true; we have been misled by the accidental circumstance that England, western Europe, and the United States happen to be places where coal is very abundant. But, relatively speaking, those regions stood just as high in civilization before the age of machinery as since. They were leaders then as now. The important factor in determining the presence of manufacturing is the character and energy of the people. If anyone doubts this, let him study the progress of manufacturing in Switzerland, Sweden, New England, Japan, or on our own Pacific coast, in spite of small supplies of coal. Hence Figure 39 (page 284), showing the percentage of the population engaged in manufacturing, is in many ways a map of the distribution of progress. It also reflects the location of certain natural resources such as cotton in the South and Cuban tobacco which is manufactured on a large scale in Florida at Tampa and Jacksonville. Nevertheless, the main aspects of Figure 39 are surprisingly like those of Figures 37 and 38. The similarity of our maps of education and manufacturing to one another and to the map of progress based on the opinions of our twenty-three contributors, is too close to be accidental. It seems to show that all three present a fairly accurate portrayal of the actual distribution of human progress. Since our two maps of climatic energy likewise not only resemble one another but are closely similar to the map of mortality, we may feel fairly sure that in all parts of the world the same relationship holds true.

CHAPTER XIII

THE CONDITIONS OF CIVILIZATION

WE have now reached a crucial point in our investigation. We must compare the distribution of civilization and of climatic energy. The reader has doubtless noticed that our maps of climatic energy in the United States, whether based on factory work or health, display an unmistakable similarity to the maps of human progress based on education, manufacturing, and general opinion. In spite of minor discrepancies, all the maps show the following features: (1) a high area elongated east and west from New England and New Jersey to a little beyond the Mississippi, (2) another and much narrower high area elongated north and south along the immediate coast of the Pacific Ocean, (3) a decline from north to south in the east, and (4) a low bay running across the country in the Rocky Mountain region, with a pronounced accentuation in the dry Southwest.

While these details are fresh in mind let us turn to three maps of Europe showing the relation of climatic energy (Figure 40), health (Figure 41), and progress (Figure 42). The map of climatic energy is based on our factory data but with some modification, because we have found that the optimum temperature for health is 64° or 65° rather than 60° as was inferred from factory work. The second map shows the distribution of health. It is based on the official mortality statistics of the European countries for the years 1909-1913. Infants under one year of age and old people over seventy-five have been omitted because the figures for those groups, especially the infants, are unreliable. The data for other ages have been reduced to what is called a standard population, so that differences in age and in the proportion of children from country to country have been eliminated. The map shows the actual distribution of health in Europe under normal conditions. The third map illustrates the distribution of civilization. It is the same as Figure 23 in the chapter on civilization, but is reproduced here in another form in order that it may easily be compared with the other two.

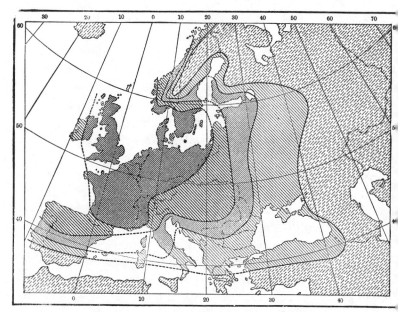

From Huntington and Williams' Business Geography, *by permission of*
John Wiley & Sons, Inc.

Figure 40.   Distribution of Climatic Energy in Europe

These three maps are so much alike that if the titles were removed, most people would not be able to tell which is which. In each map an area of heavy shading surrounds the North Sea, and shades off gradually in every direction. In each there are three projections of heavy shading, one toward Italy, a

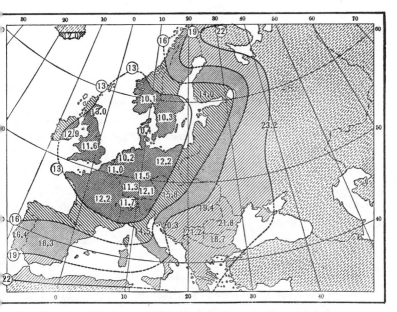

Figure 41. Distribution of Health in Europe

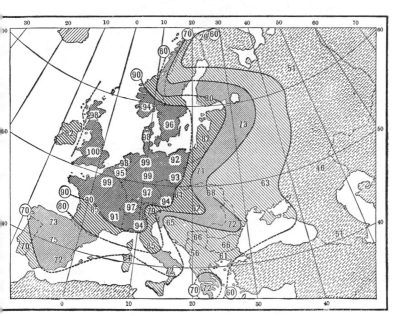

Figure 42. Distribution of Civilization in Europe

second toward the Black Sea, and a third along the Baltic. The maps are so much alike that there can scarcely be any question as to the reality of their relationship. And that relationship can be of only one kind. Civilization and progress undoubtedly influence health, and health in turn has an effect on progress. But neither civilization nor health can have any appreciable effect upon the distribution of climate. The only way in which the three maps can be so alike, unless by sheer accident, which is practically impossible, is for climate to exert a direct and dominating influence upon the distribution of health and an indirect and perhaps less dominating influence upon civilization through other agencies, such as agriculture.

Now turn to the world as a whole. Unfortunately the mortality records of many backward countries are highly inaccurate, while for vast areas they are wholly lacking. Hence it is impossible to construct a map of health. We can, however, construct maps showing the distribution of climatic energy and of civilization. A map of climatic energy based on factory work has already been given in Figure 22. This is repeated in Figure 43. The map of civilization on the same page, Figure 44, sums up the data already shown in Figures 23 to 28. All regions to which our fifty contributors assign a rank of 8.5 or higher are rated as "very high," and are shaded in solid black. Those from 7 to 8.5 are rated as "high," and are shaded in heavy lines; those from 5 to 7 are "medium," and are indicated by light lines; 3 to 5, "low," shaded with abundant dots; and under 3, "very low," and dotted only lightly.

The first thing that attracts attention is the general resemblance between the maps of energy and of civilization. Both, for example, show a high area in northwestern Europe. A tongue extends into Italy, another toward Roumania, and a third to the Baltic. Another projection runs out into western Siberia. Here the high area of the map of civilization extends about as far as the medium area of the map of energy. This is not surpris-

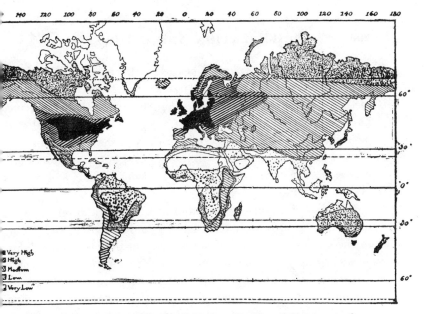

Figure 43. The Distribution of Human Health and Energy on the Basis of Climate

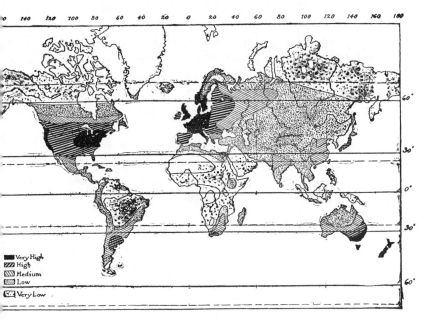

Figure 44. The Distribution of Civilization

ing, for even if the people of Siberia have the energy indicated in Figure 43, they are hampered by the remoteness and newness of their country, not to mention other conditions. In central and northern Siberia, the difference between the two maps is slight. The significant thing is that in both there is the same falling off toward the center of Asia. Still farther east in China and Japan conditions are once more alike, China being medium and Japan high.

In Indo-China and especially in India, the maps differ. Apparently, this arises largely from European domination, and is due to the constant addition of strength from that continent. This does not apply to Siam, however, which has worked out its own salvation. It ranks as very low on the energy map, and only as low on the other. This may have no significance, for our maps are still in their early stages. Further knowledge may change such slight disagreements into agreements. On the other hand, it may increase the disagreement. In that case we may discover that by long residence within the tropics, the races of Indo-China and India have become differentiated from Europeans and are less susceptible to the influence of steady heat. Again, race differs from race in its inheritance, and the Siamese may inherit stronger traits than are possessed by their neighbors. Finally, the level of Siamese civilization may have been raised by contact with other races, by the adoption of particular institutions of government, forms of religion, or social organization, or by the inspiration and energy of a few men of unusual gifts. I mention these possibilities not because they are of special importance in Siam, but because they illustrate the many and varied influences which coöperate to determine the position of a country in the scale of civilization.

In comparing the maps of energy and civilization, one of the clearest features is the effect of a strong race upon regions which it rules or colonizes. Again and again the presence of such a race causes a region to be higher in civilization than would be expected on the basis of climatic energy. Java, the Philippines, and India are examples. It is especially noticeable in regions controlled by Great Britain. In Australia, for instance, the general decrease in both civilization and energy from southeast to northwest is the same in both maps, but the presence of the English raises the places of "very low" energy to "low" in civilization, and so on, each grade being raised one degree, so to speak, until the map of civilization shows a large high area in the southeast. In South Africa and Egypt, British influence is displayed in the same way. In the Canadian Northwest, on the other hand, it is not apparent. The northern parts of Alberta and Saskatchewan appear higher in energy than in civilization. We have already seen that according to American and, to a less extent, British opinion, this is not the case, for recent settlement has raised these regions to a comparatively high degree of culture.

In the United States the energy map shows a strip of medium conditions along the southern frontier, but this is rated as high on the other map. Such a condition illustrates how a high type of government causes efficient people to settle in unfavorable regions, and how it also adds to the effectiveness of less efficient people such as Mexicans and negroes, thus in part overcoming the handicap of climate. In the central states, on the other hand, civilization is not rated so high as one would expect on the climatic basis. Probably this is because the country is so new that our Chinese, Russian, Spanish, and other foreign contributors, though they have traveled and studied extensively, do not realize how great is the progress of recent times. California, like the southern states, is higher on the map of civilization than on the other. As already explained, this may in part be due to the impossibility of making a wholly accurate map of climatic energy. It may also arise from the location of California on the seaboard, and from its early development as contrasted with the newer states of the interior. A comparison of the United States as it appears on the world-maps and as it appears on the maps of that country alone is important. Where the country stands by itself, and its parts are classified by people who live in it and are thoroughly familiar with it, the resemblance between climatic energy and civilization is greater than where the classification is on a rougher scale and is made by people less familiar with it. Another reason for the difference is that a classification of places where a uniform standard of culture prevails and where the same race is everywhere dominant is much easier than where many types of culture and highly diverse races are considered. The maps of the United States represent the kind which must be made for each country. The difference between the features of the United States on the world-map and on the other represents the extent to which our general map of civilization is in error. In spite of this, however, the general features of the country are unmistakably the same on both maps. So far as the conclusions of this volume are concerned, it makes no difference which we use. In this lies the importance of our various tests of the United States. They show that although much remains to be done before we can construct a map which is approximately perfect, the most important features are reasonably distinct and unmistakable.

Turning to Latin America, we find about what would be expected in Mexico and Central America. The highlands are medium and the lowlands low. South America, on the contrary, presents some unexpected features. The Andean highlands, including Venezuela, Colombia, Ecuador, Peru, and Bolivia, are all ranked as low in civilization, whereas the climatic map would indicate medium energy. In the belt of highlands on the east side of Africa the same phenomenon is observable. Perhaps an equatorial climate is more debilitating than would be expected from the work of factory operatives in summer. In South America the presence of an ancient race whose vigor was already waning at the time of the discovery of the New World has doubtless hindered Spanish immigrants in accomplishing what might otherwise have been looked for. This, however, does not alter the case, for the original inhabitants in the Andean countries, just as in the African highlands, stand lower than would be expected. Argentina, on the contrary, goes to the opposite extreme, and is higher on the map of civilization than on that of energy. The importance of this must not be overrated, for the climatic data are somewhat doubtful because of the paucity of statistics as to changes of temperature from day to day. As the maps now stand, however, they are encouraging, for they suggest that even with a moderately favorable climate, the Latin race in America is competent to rise to a high level.

Let us turn now from these details, and look once more at the general aspect of the two maps. In spite of minor disagreements, the main features are essentially alike. There are, in each case, the same two great high areas in western Europe and the United States. The decline from western Russia eastward to the center of Asia, and the rise to high conditions on the eastern edge of Asia in Japan are equally apparent. Likewise, the maps are strikingly alike in the shape of the very low areas in Africa and South America. South of latitude 30° S. each of the southern continents begins to rise in energy and in civilization, and the rise is more pronounced on the eastern side than on the western. Even where the maps disagree, the explanation of the disagreement is often obvious from a consideration of the recent movements of European peoples. Some of the remaining discrepancies are explicable on well-known grounds, such as the impossibility of agriculture, which hinders civilization in the far northern parts of America and Asia. In addition to all this many differences in the degree of progress among people in similar climates are due to racial inheritance. A conspicuous example of this sort, as I have shown in *The Character of Races*, is the contrast between the progressive Icelanders and the backward people who live in a similar climate in the southern tip of South

America. A still more conspicuous example is the difference between the primitive hill tribes of India and the highly competent Parsis whose ancestors came to India more than a thousand years ago because of their loyalty to their old Zoroastrian religion.

When allowance is made for obvious facts like these, the resemblance between the two maps becomes more striking. Call to mind the method of their construction. Neither represents the personal opinion or bias of any one man. Any other person with the same data before him would have obtained similar results. The maps simply give expression to two distinct sets of facts. The first is that the opinion of men of many races agrees as to the general distribution of civilization. The second is that if the various conditions of climate produced the same effect upon all the people of the world as upon students and factory operatives in the eastern United States, the amount of work accomplished in different countries would be closely proportional to the status of civilization.

Aside from the map of climatic energy, it is hard to think of any other which would so closely reproduce the features of the map of civilization. Suppose that race were made the criterion, and that a map were shaded in proportion to the number of Teutons. We should find that in Europe such a map would closely resemble the map of civilization except that places like Finland, southern France, central Italy, Hungary, Bohemia, Servia, and others are relatively high in civilization even though none are more than half Teutonic, and some only very slightly, or almost unappreciably so. In Asia, on the other hand, there is much more Teutonic blood in Syria and Asia Minor than in Japan, yet Japan ranks far higher. The Japanese might claim racial superiority almost as fairly as the Teutons, and both the Latins and Slavs may justly point to the fact that they predominate in some of the most advanced portions of the globe. When we look at the low places, we find that Teutonic areas, such as the Transvaal, Alaska, southern Greenland, and parts of Australia make a poor showing; the Latins in parts of Latin America are even worse; the Slavs at their worst fall no lower than the Teutons; while the Japanese nowhere fall so low.

A map of religion does not resemble a map of civilization, no matter which religion is employed. Protestant Christianity, indeed, prevails chiefly in regions which are either high or very high; Iceland is by no means an exception, for its civilization is much higher than would appear from our table, where it is grouped with Greenland. Roman Catholic Christianity, on the other hand, prevails in locations which range from very high to very low; and Greek Christianity from high to low. Buddhism, likewise, ranges from high in Japan to low in Tibet, while Mohammedanism never rises above medium, and in some places falls very low. That religion raises or lowers the tone of a country I do not for a moment question, but if a people are physically weak and are lacking in self-control because of something in their surroundings, the history of the world as exemplified by groups of people who have long been nominally Christians in Abyssinia, India, Latin America, and elsewhere seems to show that they debase even the finest religion. The higher the form of religion and the more self-sacrifice and devotion it requires, the more difficult it becomes to keep any but the most energetic and determined races even approximately true to it. The only cases where people of low efficiency seem to remain true to a high religion are where they are continually stimulated by the presence of a stronger race.

As a third criterion, suppose that we take form of government, and inquire whether a map of governments would resemble one of civilization. Of course, the excellence of the government is closely related to the degree of civilization, but not so with the form. Republics range from very high in Switzerland and France to very low in Venezuela. Limited, but autocratic monarchies existed in high countries like Germany, at least, before the great war, and also in low countries like Turkey and Persia. Thus we might go on to consider one after another of the great factors which coöperate in giving form to modern civilization. The nature of a nation's religious faith, its form of government, its social organization, its ease of intercourse with other nations, and various other conditions play a fundamental part in the distribution of civilization. Yet each is conditioned by the degree of energy possessed by a people, for if a race lacks energy, no amount of excellence along other lines will place it in the first rank. Energy, in turn, is greatly influenced by climate, and thus climate becomes an essential element in determining the status of civilization. We may well reverse our statement, however, and say that no amount of energy will make a nation great if none of its people are gifted with genius, or if it never evolves an orderly form of government or a moral code which allows a man to enjoy life, property, and home without constant fear of outsiders. Thus, the material and immaterial elements of civilization play into each other in such a way that either seems the more important according to the angle from which we view it.

The interplay of diverse factors is so important that it is worth while to examine it in a concrete case.* Dr. Scott Nearing in the *Popular Science Monthly* for 1914 published an article entitled, "The Geographical Distribution of American Genius." While it is impossible to measure genius, it is possible to ascertain how many people of unusual ability are born in a given region. That useful publication, *Who's Who in America*, though not infallible, forms a good summary of about twenty thousand people who have either achieved "special prominence in creditable lines of effort, making them the subjects of extensive interest, inquiry, or discussion in this country," or who occupy positions which could scarcely be attained except by persons of

---

* The following discussion is reproduced unchanged from the first edition of this book. In *The Character of Races* I have considered the changes in the past ten years, but they do not alter the conclusions set forth below.

unusual ability. Taking *Who's Who* for 1912-1913 as a basis, Nearing has tabulated the birthplaces of the first 10,000 names according to states. He took only 10,000 because that number seemed enough to give reliable results. His tabulation strikingly reënforces the common opinion that New England, especially Massachusetts, has produced far more than its proportionate share of persons of unusual ability. The utility of his investigation seems so great and the method so reliable that I have asked Dr. Nearing for permission to make use of fuller data than were contained in his article, and he has kindly supplied me with the figures for each state. In order to determine the relative status of the various parts of the country it is not fair to compare the number of eminent persons who were born in a given area with the present population. At the time when the men who are now prominent were born many of the western states contained only a handful of settlers. It is equally unfair to compare the number of such persons who live in a given region with the present population, for many persons who have achieved prominence owe it to the place where they grew up and not to that where they now live. The only fair way seems to be to ascertain the relation between the number of eminent persons born in a given region and the population of the region at the time of their birth. Accordingly, the first thing to do is to find when the people in *Who's Who* were born. Nearing gives the following table:

NUMBER OF EMINENT PERSONS IN THE UNITED STATES IN 1912
WHO WERE BORN AT CERTAIN TIMES

TIME OF BIRTH	NUMBER
Before 1850	2,818
1850-1859	2,715
1860-1869	2,717
1870-1879	1,304
1880-1889	95
1890-1899	2
Unknown	349
Total	10,000

The number who were born before 1840 is not given, but it must be considerable, for the people who attain eminence are among the most long-lived portions of the community. On the other hand, the number who were born after 1880 is too small to be considered. People rarely become eminent before they are at least thirty-five years of age. The forty years from 1835 to 1875 cover the births of practically all who had attained sufficient distinction to be included in *Who's Who* for 1912. Accordingly, we must find the average population of each state according to the censuses from 1840 to 1870, but inasmuch as the number who were born previous to 1840 is less than in later decades, we shall come nearer to the truth if we give that census only half as much weight as the others. In all cases we employ the figures for the entire white population, whether native or immigrant, but omit the negroes, Chinese, and Indians. The way in which the matter works out is illustrated in the following table, where the population is given in thousands:

	White population in thousands					Eminent persons	Eminent persons per 100,000
	1840	1850	1860	1870	Average 1840-70		
Massachusetts . . .	729	985	1221	1443	1147	1123	98.0
South Carolina . , .	259	275	291	290	282	111	39.4
Nebraska	—	—	29	122	43	24	55.8
New Mexico	—	62	83	90	67	1	1.5

The figures in the last column show the relative rank of these states in the production of persons of unusual ability from 1835 to 1875. Similar figures for each state are given in Figure 45. Since Nearing used only the first 10,000 names of American-born persons in *Who's Who*, or only about 60 per cent of the total, the index figures really mean the number of eminent persons for every 60,000 people instead of 100,000. In a few cases where the average population previous to 1875 was less than 10,000, two or more adjacent states have been combined so as to give a total large enough to be significant. The numbers thus obtained have been enclosed in parentheses. On the map the United States has been divided into four grades, much as in Figures 34 to 39. Thus all these maps are comparable. The only essential difference is that Figure 45 belongs to a period averaging more than half a century earlier than the others. It presents the most accurate picture now available of the distribution of ability at that time. New states are at no disadvantage compared with the old, for if a region had no population previous to 1860, for example, and only a few thousand in 1870, full allowance is made for this. Many of our 10,000 eminent people have moved away from their early homes, but the great majority did not go until they had at least reached an age approaching twenty and the main elements of their character were already formed. Thus the peculiarities of the map depend not only on whether the population was of such a caliber that children of high ability were produced, but also on the conditions which molded the early life of such children.

Aside from accidents three chief conditions determine the number of eminent persons in a community. The first is inherited ability. Unless a man is born with more than the average mental capacity, the chances of his inclusion in *Who's Who* are slight. The second condition is opportunity in the broadest sense of the word. A bright child born on a remote farm in Maine, on a ranch in Arizona, or in a clearing among the Tennessee mountains may be so hampered by lack of education and of the stimulus derived from contact with people outside his own little circle that he never accomplishes anything that attracts attention. The third condition is energy. Many a man of high ability, who is also blessed with the best education and with all sorts of opportunities to develop his talents, fails to make any impression on the world because he is indolent. Frequently, a man of less ability but endowed with energy achieves much more. Energy depends partly on inheritance, but also on climate. So far as it depends on inheritance it should be included under the first of our three conditions. Thus the three may be briefly defined as (1) inherited qualities of all kinds, (2) opportunities, which include education, the degree of culture in a community, and the freedom with which a person can find scope for his particular talents, and (3) energy so far as this depends upon physical circumstances not connected with either heredity or opportunity.

Let us now inspect Figure 45 to see how far our three conditions make themselves evident. Each gives rise to certain features which stand out unmistakably. To begin with inheritance, Massachusetts gave birth to 98 eminent persons for every 60,000 of its white population during the specified period. That is, 1 white child out of every 600 born at that time has distinguished himself. The figures for the surrounding New England States and New York range from 50 to 78. Such a striking difference is certainly not due to climate. It is equally certain that it is not due to opportunity. The average child in New York has as good a chance to go to school and enter any sort of occupation as has the child in Massachusetts. Yet the rank of New York is only half as high as that of Massachusetts. In Maine, Vermont, and New Hampshire the opportunities are distinctly less than in New York. There is much less wealth, the people are more isolated, the number of cities is proportionately smaller, the common school system is no better developed, and the facilities for sending children to college are not so great. Yet even Maine outranks New York, for she produced 54 eminent persons per 60,000 while New York produced only 50. Yet New York itself stands very high. Aside from the New England States only Nebraska exceeds it, while Oregon and Delaware rival it. South Carolina is another state which stands far higher than its neighbors, for although 39 is low compared with the 78 of Connecticut, for example, it is high compared with the 24 of North Carolina and the 25 of Georgia. Probably, heredity plays an important part here, as in other cases; although, as we shall shortly see, the matter is complicated by other conditions. Oregon and especially Nebraska, however, are unmistakable. Proportionately, they stand as high above their neighbors as

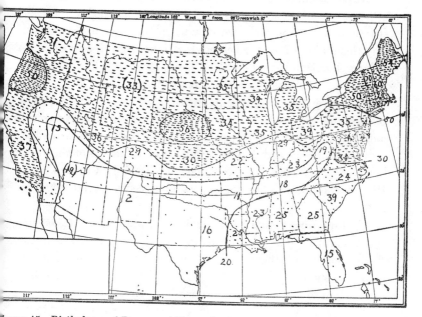

Figure 45. Birthplaces of Persons of Unusual Ability in the United States. The Numerals Indicate the Number of Eminent Persons Born in Each State per 60,000 of the Average White Population from 1835 to 1875

Massachusetts, Connecticut, Rhode Island, and Vermont above theirs. The case of these two states is most suggestive. So far as energy is concerned, there is nothing in the climate of either Oregon or Nebraska to give them a special advantage. Previous to 1890, by which time the education of four fifths of the people in *Who's Who* was completed, these two states did not offer their children especially great opportunities. In fact, the opportunities were much less than in Oregon's next neighbor, California, or in Iowa, Illinois, Indiana, Ohio, and Pennsylvania, the states directly east of Nebraska. Yet Oregon ranks 30 per cent higher than California, and Nebraska exceeds Illinois by 60 per cent, and Indiana by over 90 per cent. In striking contrast to Nebraska we find New Mexico with a rank of only 1.5, which appears as 2 on the map because we have avoided the use of fractions. Here again neither climate nor opportunities explain why this state falls so far behind its neighbors. The only reasonable explanation is that until 1870 or later its "white" population consisted almost wholly of Mexicans.

In reading the preceding pages it may have occurred to the reader that the preëminence of New England is only apparent, not real. It may be due largely to the local prejudices or limited viewpoint of the compilers of *Who's Who*. This is not the case, however. The book is edited and published in Chicago. Yet Illinois and the neighboring states all receive a relatively low rank.

The facts just stated are of profound significance. Massachusetts, because she was settled by the strong-willed Pilgrim Fathers and by other Puritans who fled to the wilderness to maintain their high ideals, has produced vastly more than her proportion of the men who have made America what it is. Connecticut and Rhode Island for similar reasons have followed closely on her heels, while the northern New England States have much more than held their own compared with the rest of the country. It has been a fad to decry puritanism, but people of puritan descent have taken the foremost place. They have done so because they inherit the strength of mind which made it possible for the Puritan Fathers to develop their stern conscientious system and carry out their noble purposes in the face of temptation and opposition. When that old stock has been transported to places such as Nebraska and Oregon, where for a while it was dominant before the great tide of later immigration, it raised the average ability to a level reached nowhere else except in New England. In New Mexico, on the contrary, we harbor a group of people, fortunately small, who are even more conspicuous by their lack of ability than the New Englanders are in the opposite way. We may excuse the Mexicans by saying that they do not learn our language and do not merge themselves in our civilization. The competent Mexicans, however, usually those who possess the greatest proportion of Spanish blood, do learn English and make themselves felt among us. The others, perhaps because they inherit an inert disposition from their Indian ancestors, are content to remain backward.

This brings up the great question of immigration and racial character. In the earliest days of colonization we received only the stronger elements of the various European populations. The North had its Pilgrims, Puritans, Quakers, and others, while in the southern states a part of the settlers were people who as Huguenots or other religious refugees were notable for tenacity of purpose and high ideals. The rest of the settlers were in large measure people of unusual courage and initiative, for others were not brave enough to come. For this reason, apparently, the states of the Atlantic coast from Georgia northward stand higher than those west of them. After America had been settled so long that migration thither was easy, we began to get immigrants of medium grade, not the best nor the worst, but from advanced countries and from the substantial middle classes. These are what predominate from Pennsylvania to Iowa. They are good material, but not so good as the old. Otherwise why should so fine a state as Wisconsin have produced only half as many eminent men per 100,000 as has Connecticut, and no more than South Carolina, which labors under far greater disadvantages? In these last decades we are taking into our midst many people scarcely better than the Mexicans. We may say what we choose about absorbing them and making them good Americans. It is our duty to do so as far as we can, but why blind ourselves to the facts of biology? Plough horses cannot race like thoroughbreds. Do men gather grapes of thorns or figs of thistles?

Today Massachusetts and New England seem to be losing their supremacy in the production of men of special ability. A study of *Who's Who* for 1922 shows that the conditions of ten years earlier still prevail, although less marked. But where the old New England families have sent their sons out over the wide expanse of our land, the loss to the mother states is more than compensated by the gain to the rest of the country. Unfortunately, the change means more than that. It means, first, that we are steadily diluting our strength. We are acting as would a dairyman who thought that by adding a dozen low-grade animals to his herd of a hundred prize-winners and letting them breed together he was going to increase the value of his stock. In addition to this we are losing in another and more dangerous way. It is as if the dairyman should not only add poor animals, but should also prevent his best animals from bearing young. No amount of care would make the low-grade animals give as much milk or be of as much value as the prize-winners. Man is subject to the same biological laws as animals. High mental ability and strength of purpose are his most valuable qualities. Yet we act as if we thought that though these are not reproduced, our country can continue to advance. Our unwillingness to live simply either prevents a large proportion of our most competent men and women from marrying, or causes many of those who marry to have few children. All men are not created equal biologically, and it is the best who are dying out. We must recognize that fact, and act upon it before we have worked irreparable injury. All this has been said many times by eugenists, but it must be repeated again and again until it is not only believed but acted upon. Biology teaches it; common sense insists upon it; and now our purely geographical studies enforce the same conclusion.

The second condition which controls the distribution of people who attain eminence is opportunity. This appears unmistakably in only one portion of Figure 45, but there it stands out sharply. Notice how West Virginia with 19 eminent persons per 100,000, Tennessee with 18, and Arkansas with 11 fall below the surrounding states. This is apparently because these are the portions of the South where mountains and other physiographic disadvantages cause the people to degenerate into "poor whites" and "crackers" in spite of a good inheritance. What these backward communities need is a "chance." They need the opportunities that are brought by schools, railroads, factories, and the other appurtenances of civilization. They need also the opportunity brought by freedom from such bodily afflictions as the hookworm disease. Kentucky, which now has a rank of 23, would probably stand much higher were not a large part of the state peopled by mountain whites. The same is true of North Carolina. Perhaps this state would not equal South Carolina, which had a large number of old families of unusual ability, but the two would be much nearer than now. A large fraction of North Carolina consists either of mountains or of swampy, unhealthy tracts along the coast, while South Carolina is almost free from such disadvantages. In another portion of the country it may be that Maine lags behind New Hampshire and Vermont in part because of her relative remoteness and lack of opportunity. Doubtless other places show the same conditions, but the matter is not certain. For example, Nevada's low position with a rank of only 15 is probably due in part to this cause, but it is doubtful whether she has been much worse off than Utah, which has the respectable rank of 36. Idaho, Wyoming, Montana, and the Dakotas certainly had no more opportunities than Nevada previous to 1890, for they were supplied with fewer railroads, and were much less easily in touch with the rest of the world. Yet their rank is 33, or more than twice that of Nevada. Taking the map as a whole it seems that although opportunity is highly important, it is less important than heredity.

The preceding paragraph stands as it was written ten years ago. Today I am inclined to think that I have not given sufficient weight to inheritance in estimating the position of the mountain whites. Note first, however, that in Figure 36, where climatic energy is estimated on the basis of the effect of climate on health, the mountain white region, by reason of its altitude, forms a peninsula of more favorable conditions jutting southward. Some non-climatic factor seems completely to overshadow this climatic condition. That this other factor is in part the isolation of the mountain valleys I do not doubt. But this is probably supplemented to a large degree by inheritance. When people are able to move freely from one region to another, there is a strong tendency for those who are most competent to gain possession of the best places. The rich cotton soil of the Black Belt of Georgia and Alabama, as R. M. Harper has well shown, has attracted and held a large number of unusually competent families—the kind that produce leaders. In such a region the lands of the less competent people are gradually bought by the competent. The less competent often move into the mountains, where land is cheap. There is likewise a backward movement whereby the most competent mountaineers drift to the lowlands. Thus in course of time there arises a genuine inherent difference of ability, and that may be one cause of the backwardness of the mountain whites.

Turning now to our last factor, that is, energy as determined by climate, we see that, in general, the outlines of Figure 45 are like those of the climatic map in Figure 36. To be sure there are important differences. For instance, the very high area which covers all the northeast and center of the country in the energy map is split into a New England and a Nebraska portion in the map of ability. Yet the Nebraska area of many eminent people displays an interesting resemblance to the tongue which projects out in the same direction on the energy map. The Pacific coast is likewise high on both maps, although there are differences of detail. Nevada, too, is at the head of a tongue of low conditions in both cases. On the Atlantic coast both maps rise from Maine to Massachusetts, and decline from New York to Florida. In the map of ability, however, high conditions go somewhat farther south than in the other map, and South Carolina, presumably because of heredity, rises unexpectedly. The other southern states from Georgia to Louisiana are also a little higher on the ability map than on its companion, probably because of the wealth and opportunities which prevailed in them previous to the Civil War, or else because of the abundance of old families with high ideals and strong minds. Yet even these states are lower than the tier of northern states from Pennsylvania to Iowa, where the average inheritance is probably no higher, if as high, but where the climate gives energy.

Taken as a whole the map of ability is an admirable example of the way in which a variety of factors coöperate in determining the status of civilization. Climate, as it were, paints a broad background, shading gradually from very high in certain areas to lower in others. Then the other factors come into play. They paint fresh colors which may or may not resemble those of climate. In some cases, such as Massachusetts, the same color is laid on by climate, heredity, and opportunity, not to mention proximity to the sea and to Europe, facilities for manufacturing, and various other factors which perhaps may be considered as opportunities. Where that happens, high civilization is sure to prevail. In other cases, such as South Carolina, the climate paints only a moderately high color, inheritance paints a higher one, education a low, the presence of the negroes a still lower, and so on indefinitely.

Such, then, is the meaning of our maps. They do not indicate that climate is the only factor in determining the condition of civilization, or even the main one. Far from it. Yet they indicate that it is as essential as any other. Today civilization seems to make great progress only where a stimulating climate exists. A high civilization may be carried from such places to others, but it makes a vigorous growth and is fruitful in new ideas only where the climate gives men energy. Elsewhere it lags, or is kept at a high pitch only by constant reënforcements from more favored regions. In the past men have perceived that climate is apparently one of the most important conditions which favor or retard the growth of civilization. They have been greatly impressed not only by its effects upon their own bodies and minds, but by the fact that in warm countries the amount of progress is closely in harmony with what would be expected on the basis of one's own feelings. At the same time they have realized that among countries located in the same latitude there are differences of culture almost as great as between temperate and tropical countries. This has seemed to indicate that climate is not so important as the tropical regions would suggest. Now, however, we see that when people's actual achievements under various climatic conditions are measured, we must revise our opinion. Variations of temperature from day to day are much more important than has been realized. Therefore, in the same latitude the stimulating effect of the climate may differ greatly. The civilization of the world varies almost precisely as we should expect if human energy were one of the essential conditions, and if energy were in large measure dependent upon climate.

CHAPTER XIV

THE SHIFTING OF CLIMATIC ZONES

WE now have before us the main hypothesis of this book, as stated at the end of the last chapter. But even if many facts suggest that civilization at present varies from place to place almost precisely as we should expect if it were dependent upon climatic energy and health, is there not abundant evidence to the contrary? Three objections at once present themselves: (1) The great nations of antiquity developed their culture in regions where the climate is now relatively unstimulating. (2) The American Indians, even when they lived in some of the world's best climates, failed to evolve a high civilization. (3) People of European races are today maintaining and developing a high civilization in relatively unstimulating climates such as that of northern Australia. Each of these objections is highly significant and must receive careful attention. The first will form the subject of the present chapter.

In a series of books, among which *The Pulse of Asia, Palestine and Its Transformation,* and *Climatic Changes* are the most important, I have set forth the hypothesis that during historic times the earth's climate has been subject to pulsatory changes. It cannot be too clearly understood that this hypothesis was not framed with reference to the hypothesis of civilization and climate presented in this book. It was developed independently before I realized how closely the distribution of civilization is bound up with that of climate. In fact, the hypothesis of climatic changes was what led originally to the studies described in this book.

The steps which have led to this hypothesis may be summed up as follows: In many parts of Asia, Africa, and America ruins of towns and cities are located where now the supply of water seems utterly inadequate. Old strands surround lakes that are now dry; old alluvial terraces in which lie remains of human occupation show that the rivers have changed their habits of erosion and deposition since men began to become civilized. Old roads traverse deserts where caravans cannot now travel; traces of dry springs are seen; bridges span channels which carry no water for years at a time; old fields are walled and terraced in places where now the rainfall is too scanty to permit agriculture and where no water can be brought for irrigation. Elsewhere old irrigation canals abound in districts where today there is little or no water supply, or where what water there may be in the streams is so salty that it kills the crops instead of invigorating them. These things and many others, which almost every traveler in semiarid or desert countries has seen for himself, seem to be almost irrefutable evidence that at some time the climate was moister than now.

Such evidence long ago gave rise to two hypotheses, which are now almost abandoned, those of deforestation and of progressive desiccation. According to supporters of the first hypothesis, the reckless cutting of forests has not only allowed the rains to denude the mountain sides of soil, but has caused an actual diminution in rainfall. This view once had a considerable popular vogue, but for various reasons it has now practically ceased to be considered among scientists. In the first place, modern measurements of rainfall before and after the deforestation of large tracts are contradictory. At best they show only slight differences, too small to have any appreciable effect, and in practically every case so doubtful that they may be due merely to the accident of an especially dry or rainy period of a few years coming not long before or after the forest was cut. Moreover, many of the strongest evidences of desiccation are found in places such as southeastern Syria where there is no reason to think that the country has ever been forested since it was first occupied by civilized man. Finally, there are thousands of square miles in Chinese Turkestan where the forests themselves have died because of lack of water, and are still standing as gaunt skeletons preserved for a thousand years or more because of the extreme dryness of the air.

The hypothesis of progressive desiccation assumes that during historic times the earth has steadily been growing drier. This was my own view when I published *Explorations in Turkestan* in 1905. Many careful students still uphold it. The majority of its supporters, however, apparently think that it needs modification along lines which will shortly appear.

While evidence of more water in the past than at present is prominent in many places, there is also much of the contrary nature, less noticeable, but no less convincing. For example, ruins are located on the floor of lakes which must have been partially dry when the structures were erected. Elsewhere one finds irrigation canals in places now so damp that their construction would seem to be a waste of energy. In north Africa and Syria huge irrigation works are located in regions of another kind, which not only are dry now, but must have been dry in the past. Otherwise the Romans would not have expended such enormous labor to get water. These things and others furnish almost irrefutable evidence that at certain periods the water supply of many semiarid regions was no greater than at present. Because such evidence is less abundant and noticeable than the other kind, the believers in progressive desiccation have overlooked it. On the other hand, other students have been so impressed by it that they have held that there have been no changes of climate during historic times, and that the fluctuations which followed the last glacial epoch came to an end before the beginning of history.

There seems only one way to reconcile these two opposing

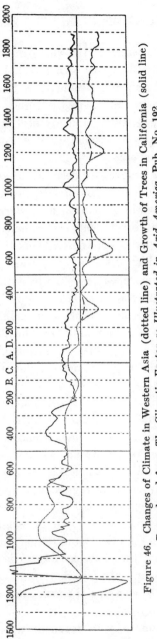

Figure 46. Changes of Climate in Western Asia (dotted line) and Growth of Trees in California (solid line) Reproduced from *The Climatic Factor as Illustrated in Arid America.* Pub. No. 192, Carnegie Institution of Washington

views, each of which is based on unassailable evidence. That way, as I first showed in *The Pulse of Asia* (1907), is to group the evidence according to its date, and see how far the indications of moisture and aridity come at different times. For example, several lines of evidence, such as the location of dwellings in tracts that are now great swamps in Ireland, unusually prolonged famines in China and western Asia, and vegetation of warmth-loving types in Europe suggest a dry period perhaps 1400 to 1200 years before Christ. At the time of Herodotus, between 400 and 500 B.C., all the evidence points to moist conditions in western Asia and northern Africa; about 200 B.C. a somewhat drier climate apparently prevailed, although not so dry as now; then at the time of Christ conditions were once more favorable. After about six centuries of gradually increasing aridity, a dry period more severe than that of 1200 B.C. reached its climax about 650 A.D. An improvement followed which culminated about 1000 A.D., then came another bad time, reaching its worst in the thirteenth century. It was followed by a rapid recovery, which did not last long enough to be of great value. Since the fourteenth century minor fluctuations have continued to take place. The whole matter is summed up in the dotted line of Figure 46. There the high parts of the curve represent moist conditions and the low dry. The curve is only approximate, and does not represent all the data now available, but it seems wise to reproduce it in the original form which it had when the "pulsatory hypothesis" of climatic changes was first formulated.

From what has been said, it appears that during historic times climatic pulsations have taken place. They seem to be of essentially the same nature as glacial epochs and post-glacial stages, the difference being only in degree. Apparently, the conditions of the geological past merge without break into those of the present. This fact is rarely appreciated by historians and archæologists, who naturally have little occasion to come in contact with it. Students of glaciation, however, have carried the matter so far that we now have abundant evidence of a great succession of climatic variations covering the entire period from the present time back to the date when ice covered much of the northern United States and Europe. They find no evidence of any sudden break between the past and the present. On the contrary, by means of "varves" or layers of clay deposited each year in lakes that disappeared after the retreat of the ice, De Geer and Antevs, for example, find a constant series of pulsations. In studying the retreat of the ice in North America, F. B. Taylor and others have discovered a series of some fifty small moraines indicating stages of retreat. These point to climatic cycles having an average duration of somewhere around five hundred years, although varying considerably. Most students of post-glacial climates believe that such cycles, on a diminishing scale, have continued into the historic period.

In order to test the pulsatory hypothesis, some means of actually measuring the climate of the past seems necessary. In the southwestern United States there appear to have been changes like those in western Asia. In that region, Prof. A. E. Douglass has found that the thickness of the annual rings of trees furnishes a reliable indication of variations in the water supply from year to year. In California among the big trees I found that correlation coefficients for a period of about fifty years show clearly that the rainfall is a main determinant of the rate at which the trees form their rings of growth. Other factors of course enter into the matter, as Antevs has pointed out in a forthcoming publication of the Carnegie Institution of Washington. But these likewise are climatic. Corrections must be made to eliminate the effects of age, but this can be done by mathematical methods of considerable accuracy. It is difficult to determine whether the climate at the beginning and end of a tree's life was the same, but it is easy to determine whether there have been pulsations while the tree was making its growth. If the trees from various parts of a given district form thick rings for a century, then thin ones for another hundred years, and again thick ones, we may be almost sure that they have lived through a long period of unfavorable climate.

During the years 1911 and 1912, under the auspices of the Carnegie Institution of Washington, I measured the thickness of the rings of growth on the stumps of about 450 Sequoia trees which had been cut for fence posts, shingles, and pencil wood in California. The trees varied from 250 to nearly 3250 years of age. The great majority were over 1000 years old, 79 over 2000, and 3 over 3000. Even where only a few trees are available, the record indicates the main fluctuations, although

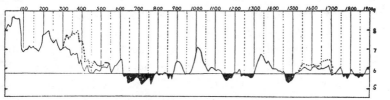

Figure 47. Changes of Climate in California for 2000 Years. Adjusted to Levels of Owens Lake. Dotted Lines Indicate Main Respects in Which Antevs' Method of Correction Gives a Different Result from That of the Author

not the details. Where the number approximates 100, accidental variations are largely eliminated. Accordingly, in California we have a climatic record which is fairly accurate for 2000 years and approximate for 1000 years more. This is expressed graphically in the solid line of Figure 46. In order to bring out the details the more reliable portion from 100 B.C. to the present time has been reproduced in Figure 47. This resembles the corresponding part of Figure 46, except that the vertical scale is three times as great, and corrections have been made on the basis of Owens Lake.

In general, the tree curve resembles that of changes of climate in Asia, although there are differences in detail. Beginning with 1000 B.C., both curves have a maximum. They dip down about 800 B.C., and rise high not far from 700. About 400 B.C. they disagree, but this is probably due largely to the absence of reliable data for the Asiatic curve. In the second century before Christ both are low, but not so low as at present; at the time of Christ they rise high, and continue to fluctuate together till 300 A.D. At that time they show a difference which may have arisen because I was unduly impressed by the abandonment of many ruins in Chinese Turkestan at the end of the third century. The dash line is probably more accurate than the dotted and should be used. In the seventh century both curves reach their lowest point. Then, from 650 A.D. onward, their general course is closely similar, especially about 1000 A.D. In the tree curve the evidence of pulsations is even clearer than in the Asiatic curve. The general agreement between the two curves seems to indicate that the main climatic variations of western Asia and the region of similar climate in the United States are approximately the same. Moreover, the fluctuations of both tree growth and rainfall among the Sierras during the last half century show a strong resemblance to those of rainfall in Palestine, as is set forth in *Climatic Changes* and in *Post-glacial Climatic Cycles*, a publication of the Carnegie Institution of Washington. Since the present growth of trees in California shows so close an agreement with the rainfall in western Asia, it is almost certain that a similar agreement existed in the past. The growth of the trees, with its pronounced variations lasting hundreds of years, is generally accepted as the strongest single piece of evidence that pronounced climatic pulsations have taken place in historic times. The agreement between the curve obtained from exact measurements in California and that obtained from physiographic and archæological evidence in western Asia unites with present climatic variations in proving that these two regions of similar climatic type have experienced a similar series of climatic pulsations.

Another line of evidence in the western United States is peculiarly important because it employs a method absolutely different from those already mentioned, but reaches the same conclusion. Every river holds in solution a certain amount of sodium, chlorine, calcium, carbon dioxide, and various other materials. Under ordinary circumstances this cannot be detected except by chemical analysis. If the river flows into a lake which has no outlet, however, the water is evaporated, but the dissolved constituents remain, and gradually increase until a brine is formed. Certain materials, such as the calcite, which forms tufa or other kinds of limestone, are removed by algæ or bacteria, and certain others, such as potassium, seem to be absorbed by the clays of the lake bottoms. Sodium, and especially chlorine, however, do not appear to be removed until the brine becomes completely saturated so that crystals of common salt are formed. Hence, if we know the amount of sodium and chlorine brought in by the rivers each year and the amount dissolved in an unsaturated lake, we can easily calculate the time that has elapsed either since the lake was first formed, or since it last overflowed. A body of water that overflows, as everyone knows, soon becomes essentially as fresh as the rivers that supply it.

On the eastern side of the Sierra Nevada mountains Owens Lake, a body of salt water in southern California, is peculiarly well adapted to our present purpose. Owens Lake receives most of its water from the river of the same name. Both have been measured and analyzed with unusual thoroughness because part of the river is carried to Los Angeles in a remarkable aqueduct nearly 250 miles long. H. S. Gale of the United States Geological Survey has carefully gathered all the available data. He concludes that, according to the figures of the aqueduct engineers, the accumulation of the chlorine of Owens Lake would require 4200 years, and the sodium 3500, the average being 3850. A series of fresh strands and an old outlet channel show clearly that the lake overflowed not long ago. Hence, Gale concludes that "4000 years or considerably less" is the length of time since the lake stood at the outlet level, 180 or 190 feet higher than at present. These figures, however, require modification. The reason why Gale gives the period as "4000 years or *considerably less*" is that the figures of the engineers omit the lower third of the drainage area of Owens Lake, and this is the part where the waters flow most slowly and where the clays and other deposits which surround them are most saline. Hence, more salt proportionately should be accumulated here than higher up. Moreover, as he carefully points out, no allowance is made for the well-ascertained fact that when the rivers are more abundantly supplied with water, as must have been the case when the lake was full, the amount of dissolved salt is also greater although not in direct ratio to the water supply. When due allowance is made for these conditions and for others of minor importance, the time since the last overflow is reduced to 2500 or 2000 years. In other words, at the time of Herodotus and perhaps at the beginning of the Christian era, the climate of the Owens Lake region was so moist that the lake expanded to two and one half times its present size and sent a stream down the outlet channel.

Owens Lake does more than indicate a change of climate within two or three thousand years. It also shows that the change has been highly irregular. This is proved by a large number of strands lying below the level of the outlets, and by the way in which these vary in character and in the extent to which they have been covered by fresh detritus washed down from the mountains. At Owens Lake there are four series of strands. These apparently correspond to the four chief periods when the climate has grown moist as shown by the growth of the big trees in Figure 47. Fortunately, Owens Lake lies only fifty miles east of the region where the trees were measured. The general climatic fluctuations of both districts are the same. The uppermost strand, the huge gravel beach at the level of the outlet, was presumably last reached by the water about the time of Christ, or possibly in the days of Herodotus, for both the chemical evidence and the trees point to this conclusion. A series of similar, but much smaller beaches at lower levels record the approach of a dry period during which the lake fell to a low level whose exact position cannot be determined. Judging by the trees this must have culminated about 650 A.D. During this period gravels were washed in by mountain streams and formed what are known as fans, or low, flattened cones, which may be several miles long. These covered the old strands in many places, and extended far below their level to the diminished lake.

Next the waters rose again, but not halfway to their former level. They formed two small strands, not gravelly like their predecessors, but faint and sandy as if the winds were weak. They must date from about 1000 A.D., when the trees indicate a wet period, for they are younger than the gravel fans of the preceding dry time. The next phase of the lake was a dry period, which was most extreme about 1250 A.D. More gravels were then deposited, and the fact that they cover the preceding strands and extend to a much lower level shows that the lake then stood low, as would be expected from the trees.

The next high period of the lake, about 1350 A.D. according to the trees, is unusually interesting. The water did not reach so high a level as formerly, perhaps because the rainy period was short, but it formed a large beach of gravel quite different from the preceding beaches. This seems to indicate great storminess, a condition which is also suggested by the fact that the growth of the trees at this time increased more rapidly than at any other period for nearly 3000 years. In Europe during the same century, unprecedented storms caused great floods in France, while the severity of the waves was so intense as to break through beaches and sand dunes, and convert large marshy areas into portions of the sea along the coasts of Holland and Lincolnshire. During the winters the rivers froze to an unheard-of degree, and three or four times men and animals passed from Germany to Sweden on the solid ice of the Baltic Sea, an occurrence unknown in our day. In England the summers were so rainy that the average yield of grain diminished disastrously. In self-defense many landowners gave up grain-raising, and turned their attention to sheep and cattle. Distress and discontent were the inevitable result among the peasants. Far away in central Asia the Caspian Sea and the lake of Lop Nor both rose with great rapidity between 1300 and 1350 A.D. Thus from California to China evidence of various kinds unites to indicate that during the fourteenth century there occurred a short period of unusual storminess. Such conditions, if intensified and prolonged, would probably cause the accumulation of enormous glaciers.

To return to Owens Lake, the lowest series of strands is sandy and small compared with the large gravel bar of 1350, and was evidently formed under different conditions. Presumably the lake fell to a low level about 1500 A.D., and rose during the next century or more to form the highest strand of the latest series. The evidence of Owens Lake is much strengthened by that of its neighbors. At Mono Lake farther north in California, the more recent strands are almost identical with those at Owens. Since 1900 Mono Lake appears to have risen higher than at any time since at least 1775, as we judge from the rings of growth of a tree killed by the rising salt water. Thus on all sides there is the strongest evidence not only that the climate of the past differed from that of the present, but that many minor pulsations have occurred, and have grown less intense during the historic period.

Up to about 1912 I supposed, as did practically all students of the subject, that the same kind of climatic changes have taken place in all parts of the world. At that time Penck, a leading student of the glacial period, came to the conclusion that this is by no means the case. He showed that on the north-

ern side of arid or desert areas we find densely saline lakes, like
Great Salt Lake in Utah, surrounded by old strands. These
indicate that the lakes have long been contracting, so that they
have abandoned first one strand and then another, while at the
same time their water has become more and more highly concentrated. On the equatorial side of the desert belt, on the contrary,
we have such lakes as Chad in Africa, a shallow sheet of water,
only slightly salty and not surrounded by any great series of
strands. It has the appearance of being a new lake formed by a
recent increase in rainfall. Penck points out a similar contrast
between the sand dunes on the poleward and equatorward sides
of the desert belt in both hemispheres. On the poleward side, the
dunes consist of loose, moving sand appropriate to places that
are growing drier. On the other side there are plenty of dunes,
but they are covered with a sparse vegetation which is sufficient
to keep them from moving and to prevent the formation of new
dunes. Since their formation, the rainfall has evidently increased. From the dunes and lakes, as well as from other evidence, such as the snowline, Penck concludes that changes of
climate consist of an alternate shifting of the climatic zones.
During a glacial period he holds that the northern storm belt is
shifted southward so that the storminess of Germany and the
northern United States is pushed into Italy or the southern
United States. In the same way the desert belt is displaced
toward the equator. Thus the polar side of the desert has
more storminess and moisture than formerly, while on the equatorial side the desert is shoved into an area where equatorial
rains formerly supported abundant vegetation. During a time
such as the present, on the contrary, the desert expands on its
northern border. Its lakes diminish, leaving strands behind them
and becoming very saline; vegetation dies; and the wind is free
to pile up sand dunes. On the other, or equatorial side of the
desert, the amount of rain is greater than before. Hence, basins
which formerly contained no water are now filled with shallow

lakes, such as Chad, which have not yet had time to become highly saline. Vegetation also spreads into the desert and sand dunes become covered with it and cease to be moved by the wind. Few people have studied glacial problems more carefully than Penck. One result of the minuteness and care of his studies is that he is one of the foremost advocates of great climatic complexity. He distinguishes several post-glacial stages, and in addition brings his conclusions so far toward our own time that he states that a peculiarly dry period prevailed in central Asia as late as the early part of our era.

While Penck was formulating his hypothesis of a shifting of climatic zones as an explanation of the glacial period, I was working out the same hypothesis in respect to historic times. Our work was wholly independent, being based on different lines of evidence, and, so far as I am aware, neither knew what the other was doing. The shifting of climatic zones is by no means a new idea, for it has been vaguely suggested many times. The new thing is to find direct evidence of it, such as Penck presents in his discussion of lakes and dunes. As to historic times, the ruins of Guatemala and Yucatan furnish perhaps the best available test, for those regions are the most notable example of a civilization which developed within the torrid zone. The ruins of that civilization are the most remarkable archæological remains in the western hemisphere. Part lie near the northern coast of Yucatan in a relatively dry region inhabited today by a fairly prosperous agricultural population. Many of the finest ruins, however, and most of the more ancient ones, lie back from the coast in a wilderness of dense forest and jungle. There agriculture is almost impossible. The difficulty of clearing the rank vegetation and getting it dry enough to burn before a new crop of lusty bushes grows up is enormous. Fevers, too, prevail most of the year. They include the worst types of tropical malaria, as well as many other kinds. Foreigners are quickly attacked, and it would be dangerous for a white man to attempt to live there permanently, even with all the appliances of modern medical science. The natives also suffer terribly. Many of the children are apparently killed by malaria and other tropical diseases in infancy. Those who grow up carry the effects with them through life. Seventy-five per cent of the tropical workers at the Panama Canal are thought to have had malaria germs in their systems, even though they did not show outward signs of the disease, and the case is probably even worse among the natives of the lowland forests of Quintana Roo, Campeche, and Peten, where most of the early Maya ruins are located. One has only to look at their swollen paunches to see that something is the matter; and their dull, apathetic manner, both in work and play, is another sign of some deep-seated physical ailment.

Here, where today the Indians are so diseased and agriculture is so difficult, there dwelt an ancient race characterized by the qualities which we have defined as most essential to a high civilization. The Mayas of Yucatan and Guatemala, alone among the aborigines of America, carried to high perfection the arts of sculpture and architecture. The Incas of Peru, to be sure, made striking buildings, but they had no idea how to adorn their simple structures with columns, rosettes, arches, gargoyles, pediments, and many other architectural devices which were universal among the Mayas. These clever people were also adepts in astronomy, for they framed the most exact calendar ever known except our own. Theirs was better than that which eastern Europe used up to the time of the Great War. Most remarkable of all, the Mayas developed the art of writing and carried it to a pitch higher than that reached by the Chinese, for they apparently began to use signs to denote sounds instead of having a sign for each individual word. Their temples are great structures which sometimes rise three stories in height and have a length of three or four hundred feet. A single city often contained a score of noble buildings and extended over many square miles of territory. The Mayas need not shrink from com-

parison with any people who had no greater opportunities. They
were terribly hampered in many ways. They were not blessed
with progressive neighbors to spur them on and teach them new
ways. They had no beasts of burden. They could not plough the
fields, nor transport loads except by their own labor. Every one
of the great stones which form their temples and palaces must
have been brought by human effort, a task which none but a
most energetic race would undertake. Another disadvantage was
the absence of metal tools. The Mayas perhaps had a little
copper, but not a trace has been found of any metal tool that
would be of service in carving their intricate façades and deli-
cately wrought statues. Flint or obsidian was all that they had,
and yet with such poor tools they created works of art which
command sincere admiration. To do all this with such small
opportunities was surely one of the greatest feats ever achieved
by any race.

The most surprising thing about the Mayas is that they
developed their high civilization in what are now the hot, damp,
malarial lowlands where agriculture is practically impossible.
A hundred miles away on the coasts of Yucatan or in the Guatemalan highlands far more favorable conditions now prevail.
There agriculture is comparatively easy, the climate, while not
bracing, is at least good for the torrid zone, and malarial fevers
are rare. Today the main cities lie in these more favorable regions ; the energetic part of the population is there, and the interior lowlands are hated and shunned by all except a degraded
handful. In the past the more favorable localities were occupied
by people close akin to the Mayas, yet civilization never rose to
any great height. Ruins are found there, but they are as far
behind those of the lowlands as the cities of Yucatan are today
behind those of the United States.

In explanation of these peculiar conditions several possibilities suggest themselves. First, we may suppose that the
Mayas were the most remarkable people who ever lived. They

were able to carry on agriculture under conditions with which
no modern people, not even those of European race, have ever
succeeded in coping. They chose the worst place they could
find, almost the worst in any part of America, even though far
better places lay close at hand and were occupied by an allied
people relatively few in numbers and backward in civilization.
They were able to preserve their energy for a thousand years or
more under the most debilitating climatic conditions; and, lastly,
they were immune to the many fevers which today weaken the
dwellers in their old habitat. It is possible that a process of
natural selection had given these people an extraordinary degree
of genius, energy, and ability to combat disease. Indeed, it seems
to me that this must have been the case. Nevertheless, I doubt
whether it fully explains why the Mayas settled and thrived in
what now would generally be accounted the worst parts of their
general region. A second possibility is that in the time of the
Mayas tropical diseases were less harmful than at present. We
have no shred of evidence one way or the other. Such a thing is
possible, but in view of the fact that there are several kinds of
malaria, not to mention other diseases, any one of which greatly
weakens a race, there is little chance that the Mayas were free
from disease unless something else was also different. Even if
diseases were not so prevalent as now, this does not explain the
other apparent anomalies of the Maya situation. It may be,
however, that a fairly satisfactory explanation will be found if
the two preceding possibilities are joined with a third, namely,
a climatic change such that the dry conditions which prevail a
little farther north prevailed in the Maya region when these
people attained eminence. Such a shifting of zones would increase
the length of the dry season which now comes in February and
March. This would diminish the amount of vegetation and cause
scrub to take the place of dense forest. Under such conditions
agriculture would become comparatively easy. Fevers would
also greatly diminish, for in the drier parts of Yucatan they

are today relatively mild, and the lowland plain would be the natural site of the chief development of civilization just as is the case in other countries.

When the dates of Maya history are compared with the curve of tree growth in California, they seem to agree with the hypothesis of a shifting of climatic zones. The early chronology is so doubtful that we may pass it by. The seventh century, however, is known to have been a time when Maya civilization sank to a very low ebb, for scarcely a building dates from that time, and the traditions become most vague. The great fact of this period is that for some reason the Mayas completely abandoned their old homes in what is now the forested south of Mayaland. They migrated north to the drier, less forested, and more healthful region of northern Yucatan. This would be what we should expect, for when the California trees grew slowly, as in the seventh century, the desert zone that almost girdles the globe would lie far to the north. At the same time the equatorial zone of rains would expand northward over Guatemala and Yucatan; the rainfall would be abundant and the dry season short; the forests would become rank, agriculture would be difficult; disease would be rife; and the vitality of the Mayas would be sapped. From about 900 A.D. to 1100 A.D., on the other hand, the California trees grew rapidly. At such a time the desert belt would be pushed south, and favorable conditions would prevail in the home of the Mayas. At that time occurred the last great revival of architecture and the construction of the great buildings whose ruins now adorn Yucatan. Whatever the cause, there is no doubt that there was a marked outburst of energy. The later history of the Mayas has no great abundance of architectural monuments to serve as landmarks, and hence cannot be correlated with the fluctuations of the California curve. Yet, so far as the indications enable us to judge, the two sets of phenomena are in harmony.

In addition to the Maya ruins, the conditions of Palestine

and the growth of the California trees furnish independent evidence that the zone of cyclonic storms has at certain periods suffered a shift equatorward. In Palestine, the rainy zone apparently once extended at least fifty miles south of its present limit, in spite of the fact that the mountains there die out and thus render the conditions doubly unfavorable to rainfall. Today Hebron is the last large town and Beersheba the last village. In the old days, Aujeh, far to the south, was comparable to Hebron, and other ruins indicate that settlements were located still farther southward. The California trees indicate a shifting of zones, because they grow most rapidly during years when storms continue late into the spring. At such times the winter zone of storms is pushed equatorward and continues to give moisture late in the spring.

Further consideration of changes of climate is not here possible. For that the reader must turn to *Climatic Changes* and *Earth and Sun*. Here our main interest centers in the results of such changes, and in the degree of certainty with which we can make use of them in the interpretation of history. For our present purpose the essential point is that a series of climatic pulsations has apparently continued from a maximum phase in the last glacial epoch down to a minimum phase today. Nowhere do we find any marked break between the past and the present. On the whole, the early milleniums of the historic period appear to have been drier than the present in such regions as Syria and California, but this general condition of more abundant rain appears to have been broken by frequent dry periods which may have lasted hundreds of years. Some of these, as in the seventh century of the Christian era, may have been even drier than the present. Moreover, the effect of any given phase of a climatic pulsation is not the same in all parts of the world. On the contrary, the effects differ from zone to zone, from the coast of a continent to the interior, and from the east side to the west. Except in magnitude the climatic cycles of the past appear to have been like the small cycles of the present. California, for example, tends to be rainy when Florida is dry.

It must be clearly understood that the chief characteristic of the climatic pulsations of the historic period has not been changes of temperature. Failure to understand this point has led to many mistakes. The chief variations have apparently occurred in the number, intensity, and paths of the kind of cyclonic storms which bring ordinary changes of weather in the progressive regions of the United States and Europe. Such storms are by far the most variable climatic factor. The extent to which the mean annual temperature or relative humidity of a region may vary from one year to another is very limited, but the rainfall, and still more the number of storm tracks whose centers pass near a given place, may vary greatly. Some years storm after storm sweeps across the country at intervals of only a few days, while in others a real storm occurs scarcely more than once a month. Even in a place like New York City where the climate is very regular, the number of storm tracks whose centers passed within 200 miles of the city in the short period from 1900 to 1915 varied from 21 in 1903 to 48 in 1915, while from July to December the variation was from 7 to 24. In regions having the type of climate now found in the eastern Mediterranean region and in California, such variations are far greater. At San Francisco, for example, the number of storm centers passing within 200 miles of the city during this same period varied from one in 1900, 1904, and 1914, or none in the twelve months from December, 1913, to November, 1914, to 13 in 1903. In the past, according to our hypothesis, the number of storm tracks for century after century approached this higher figure, while at other periods it fell toward the lower figures. In other latitudes and in parts of the continents where other types of climate prevailed, an opposite type of change took place, while in intermediate regions no distinct alternation was evident.

CHAPTER XV

THE PULSATORY HYPOTHESIS AND ITS CRITICS

C HANGES of climate are so vital an element in the main
conclusions of this book, that I shall devote a chapter to
criticisms of climatic hypotheses. I shall begin by summarizing
once more the changes in my own opinions, for these changes
have had much to do with the nature of the criticisms.

1. When I first became convinced that climatic changes had
taken place in historic times in western Asia I accepted without
question an opinion then somewhat widely held that the earth
as a whole was growing drier and warmer.

2. Further study, as already explained, soon led me to aban-
don this idea of progressive desiccation in favor of the hypothesis of pulsations whereby the climate first became warmer and
drier, and then moister and cooler.

3. The next step was the conclusion that variations in temperature are of negligible importance, and that the main change
has been in the amount of rainfall.

4. Investigations in regions other than deserts advanced the
hypothesis another step by leading to the conclusion that
changes of opposite types have taken place in different latitudes,
and even on opposite sides of a continent in the same latitude.

5. The last step was the conviction that the evidence points
clearly to changes in the location, number, and intensity of
cyclonic storms as the main factor in climatic pulsations.

Bearing in mind, then, that the hypothesis of climatic pulsations as presented in this book has become quite different from
what it was when first published in 1905, let us consider some
of the main criticisms.

1. Many authors have endeavored to disprove all theories of climatic change during historic times by suggesting that the supposed evidences of moisture in the past are purely accidental. In one place, so they say, a river *may* have changed its course; in another an earthquake *may* have diverted an underground water supply; famous marches like that of Alexander with his elephants in Persia *may* have been possible not because the climate was moister than that of today, but because the march occurred in an unusually wet year. In the same way the thousands of seemingly waterless ruins in the southwestern United States may all have depended upon hidden springs, one or two of which have been found. Such arguments are of value merely as guides in determining the kind of investigations that shall be made in the field. They break down when it is realized that although many conscientious investigators have had them in mind, the evidence that would give them weight is not forthcoming. It must be remembered that the evidence on which the hypothesis of climatic pulsations is founded is of many types; it is widely spread in at least four continents; and it depends upon literally thousands of ruined towns, waterless roads, abandoned irrigation systems, old unirrigated fields now too dry for cultivation, old records of lake levels and the like. It is statistically impossible that so many accidents should all point in the same direction. Accidents by their very nature tend sometimes in one direction and sometimes in another. The evidence of climatic changes, however, is not only widespread, but extremely systematic. At certain periods and in certain types of climate it all, so far as I can see, tends to indicate aridity. In other types of climate and at other periods it seems to point with equal unanimity in the other direction. If the evidence were merely the result of accidents, such a separation on the basis of both time and place would be impossible.

2. A number of persons have criticized the pulsatory hypothesis under a misapprehension, or have directed their criti-

cisms at earlier statements of the hypothesis without knowing of later modifications. A good instance of this is Woeikof's book on Turkestan. Apparently that Russian author had not read or had failed to understand *The Pulse of Asia*, but had become thoroughly familiar with a book written by his countryman, Berg, for the express purpose of combating what he believed to be the false conclusions of *The Pulse of Asia*. At any rate, Woeikof pronounced the conclusions of that book "inane," and then proceeds to prove the very thing that led to the choice of its title, namely, that during historical times the Caspian Sea and other lakes in central Asia indicate alternating periods of more and of less rainfall than at present. Herbette, a French critic, also aims his main arguments at *progressive* climatic changes, but agrees that the Caspian Sea affords unmistakable evidence of climatic pulsations. In general, the critics have tended to attack not only the idea of a progressive climatic change in one direction during historic times, but the idea of climatic uniformity. The alternative to which almost all authorities turn is climatic pulsations having a greater intensity in the past than at present. Over nine tenths of the geographers of America, if we may judge from the fifty whose opinions have been expressed in writing, hold this view, although they differ as to the degree to which climatic pulsations have influenced history.

3. A certain type of critic has asserted again and again that in the southwestern United States, and to a less degree in other places, the population in regions that are now too dry to support many people was never so dense as the ruins seem to indicate. What really happened, they say, is that the same people built village after village and house after house. Superstitious fears of death or disease, or a tendency to wander may have caused a house to be inhabited for a few decades and then abandoned. The old house never was reoccupied, but a new one was built by its side. Thus ruins which appear to be large enough

for a thousand people may never have been occupied by more than fifty or one hundred at any one time. Wherever the history of a region is well known this explanation is wholly out of the question, as in Syria, the Libyan desert, Persia, and many other parts of the Old World. It has flourished in America simply because the history of pre-Columbian America is unwritten, and archæologists have been free to exercise their imaginations. Douglass, however, who perfected the methods of measuring climate by means of tree growth, has most effectively proved that in at least one case a relatively large set of ruins was all built at practically the same time. By comparing the rings of different trees he is able to tell just which ring in one tree is synchronous with a given ring in another. This is possible because the rings vary greatly from year to year, and the same succession of rings of various thicknesses is practically never repeated. By this method Douglass finds that in one of the large ruins at Chaco Canyon, New Mexico, the houses were all constructed within about a decade. Thus the whole village must have been occupied at one time and there was no successive addition of new rooms to replace those no longer occupied. Of course this is only one instance, but a single concrete, well-established fact like this is worth more than any amount of supposition.

4. In many quarters there has been an assumption that climatic uniformity is a normal condition. Meteorologists do indeed find that so far as records are yet available, which means for scarcely a century, there are no certain indications of progressive climatic changes. But there is overwhelming evidence of pulsations which increase in magnitude and duration as the records become longer. In the same way the geological record continually discloses new evidence of climatic pulsations. Where one glacial period consisting of a single epoch was generally recognized half a century ago, at least six are now recognized. These date from all parts of geological time, and some have several epochs and subepochs. Moreover, the studies of Bar-

rell, De Geer, Sayles, and many others are rapidly showing that during practically the whole of the earth's known history smaller climatic variations have occurred. Thus pulsations of climate are now almost universally regarded as characteristic of the vast periods covered by geology as well as of the short period covered by exact climatic records. The presumption is that similar pulsations of intermediate magnitude were characteristic of the intervening historical period.

One of the most interesting facts in this connection is the change of attitude among meteorologists. For example, Ward of Harvard, who was for many years the only professor of climatology in an American university, has long been one of the strongest champions of climatic uniformity. In his later works, however, and especially in his forthcoming *Climate of the United States*, his attitude toward the hypothesis of pulsatory climatic changes has altered from pronounced opposition to benevolent neutrality. His position is that the kind of physiographic and archæological evidence which must be used in solving the problem of climatic changes during the period prior to exact records is utterly different from the statistical data to which climatologists and meteorologists are accustomed. Therefore specialists of those types cannot properly evaluate it, but must wait for the final decision of specialists in this particular line. As a matter of fact, the great majority of specialists in post-glacial climatology now accept the idea that climatic pulsations have continued into historic times.

5. The most important criticism of my climatic hypotheses that has yet appeared was written by J. W. Gregory,* a British geographer who has traveled very widely. He makes two important points, but both pertain to matters where I had already come to the selfsame conclusion which he advocates. His first point is that various types of vegetation, such as the palm tree, grew in places like Palestine in the past and grow there today.

* "Is the Earth Drying Up?" *Geographical Journal,* vol. 43, 1914.

This, he says, precludes a change of climate, for if Palestine were colder two thousand years ago than now, the palm could not have flourished. The truth of this statement is undeniable if we are talking about pronounced changes in the mean temperature. As I have already shown, however, the evidence points toward changes of precipitation much more than of temperature. The best students of glaciation agree that during the glacial period the mean temperature of the earth as a whole was probably not more than 10° or 15° F. cooler than at present. If the climate at the time of Christ differed from that of the present time by one tenth of the difference between our modern climate and that of the glacial period, the temperature of Palestine would only have to be about 1° F. cooler than now. That would scarcely produce any appreciable effect upon vegetation. The actual figures show that Palestine might be at least 2° or 3° F. colder than now without preventing the growth of the palm tree. Moreover, the extremes may have been greater than at present without much change in the average temperature. If the winter storms were colder than now by a few degrees, that would not hurt the palm tree. In Seistan in eastern Persia I have camped in six or eight inches of snow almost under the shade of a palm grove whose dates, packed in a sheepskin, were uncommonly delicious. The thermometer fell below 20° F., and a strong north wind overturned the tent, but the palm trees did not suffer, for in two or three days the air was again as balmy as spring. So far as has yet been shown, the conditions of vegetation nowhere seem to be out of harmony with our hypothesis.

Gregory's other point, like the one already discussed, agrees completely with a conclusion to which I had been led independently. He shows that so far as there is evidence of climatic changes, it points in one direction in some regions and in the opposite direction elsewhere. He even published a map showing the distribution of positive and negative changes. Inasmuch as this conclusion agrees with that to which both Penck and myself had been led on entirely different evidence at essentially the same time, I see no reason to question it. In a word, the article by Gregory, which some historians have regarded as demonstrating that there have been no changes of climate, goes far toward substantiating the hypothesis of this book, for it helps to clear up some of the main difficulties. It does not touch at all upon the problem of pulsations. Like practically every other important criticism of my conclusions it was published while the pulsatory hypothesis was still being developed and before the most convincing evidence, that of the Big Trees, had been published in *The Climatic Factor*. Since the publication of that book, so far as I am aware, there has been no serious quantitative attempt to disapprove the general hypothesis of climatic pulsations during historic times.

6. The contrast in the types of evidence which are conclusive to workers in diverse lines of research is illustrated not only by the fact that meteorologists are slow to accept the evidence of historic changes of climate, while geologists are quite ready to accept it, but by the fact that historians seem to think that if changes of climate have occurred there ought to be written records to this effect. This may be illustrated by two eminent geographers who follow the historical method. Brunhes and Semple both reject the idea that climatic changes have had an important effect upon history, but neither gives reasons for such rejection. Miss Semple, for example, in an admirable paper on the Mediterranean lumber trade merely states that, having studied the evidence, she does not think it convincing. Her work, with its remarkably complete references even when minor matters are discussed, shows that to her the real evidence on this point is the written word. With her usual fairness, however, she points out that the accounts of widespread forests in ancient Cyprus suggest a different climate. She also gives many other facts which seem much more consistent with climatic changes than with uniformity. At least, they seem so to persons with a geo-

logical training. Such persons have become more and more accustomed not only to the idea of repeated climatic fluctuations of all grades throughout geological times, but to the idea that such fluctuations have caused the destruction of many types of plants and animals, and have hastened the development and migration of new types. The historian, on the contrary, is accustomed to attribute almost everything to purely human action. Hence, in the absence of records to the contrary, he assumes that human causes are mainly responsible for such facts as deforestation, changes in trade routes, the apparent drying up of sources of water, the abandonment of settlements in the drier parts of the world, and the migrations which so largely originate in dry regions.

7. Closely connected with the preceding attitude of mind is a reason for the rejection of the hypothesis of climatic changes which has been advanced by some historians. They say that such changes are not needed in order to explain the historic facts. But this has nothing to do with the matter. The primary question is whether climatic changes have taken place during historic times. If they have taken place, the historian must perforce take note of them and inquire into their effect, just as he inquires into the effect of barriers like the Alps or of great men like Socrates. This wholesome attitude is beginning to prevail in many quarters and will doubtless grow.

A good example of the way in which a knowledge of climatic changes throws light on history is set forth by A. T. Clay in an article on "The So-called Fertile Crescent and Desert Bay."* From a study of Babylonian tablets Clay discovered that a Semitic people known as the Amaru or Amorites had already risen to a position of relatively high civilization in the fourth millennium B.C. Their capital was at Mari on the Euphrates River not far from 37° N. in the very heart of what the oriental historian Breasted has described as a "*desert* bay." The tablets

* *Journal American Oriental Society,* 1924.

seemed to show that this capital was powerful enough to rule Babylonia, but Breasted maintained that this was impossible because there was no sufficient "material basis" for an empire such as that of the Amorites. The arable land along the Euphrates is too limited to support an empire, and the surrounding land is desert. Nevertheless Legrain of the University of Pennsylvania found a fragment of a dynastic list which showed beyond question that the Amorite city of Mari on the Euphrates ruled Babylonia in the fourth millennium B.C. Thus the historic facts and the present condition of climate are completely out of harmony. When Clay investigated the matter on the spot, he found about seven hundred square miles of arable land within the alluvial tract which might be watered from the Euphrates. Much of this, however, could not now be irrigated under the present conditions of floods and erosion. Even if it could all be irrigated it is scattered in so many small bits that it would form a very inadequate basis for an empire which could rule Babylonia. The region round about, however, is full of evidence that the country was by no means a desert in the past. As Clay puts it: "To include the Euphrates valley in the so-called 'desert bay' would be equivalent to including the Nile Valley in the Sahara desert, the difference, however, being that unlike the Sahara, Mesopotamia was not a desert in ancient times.

"It is not improbable, since we know that the climate has changed within the past two thousand years, that practically the entire area of many thousand square miles of Mesopotamia north of the [Euphrates] river was fertile, or at least was a great pastoral territory.

"The Khabur and the Balikh rivers, which flowed through this region southward to the Euphrates, were augmented by numerous streams, some of these at present containing water, while others are dry. Today the Balikh river at the end of summer is almost dry at its mouth. In commenting upon the Khabur and its tributaries an Arab writer says it is 'such as is not to be found in all the land of the Moslems for there are more than three hundred pure running fountains.' Certainly the Hebrew writers and redactors of the old Testament [would not have made] themselves ridiculous in the eyes of their intelligent contemporaries by placing the Garden of Eden in this region, if it were a 'desert bay.'

"And what is true of the region north of the Euphrates, is true of the land lying to the west of the river. There are dry beds of rivers and streams with sand and pebble bottoms in which, at present, water is not seen from one end of the year to the other; even many of these streams are spanned by well-constructed bridges. There are well-heads, spring houses, in which water formerly gushed from the earth, some even containing inscriptions; but where neither wells nor springs exist today. Even in the flat and once fertile plateau, which was thickly inhabited, there are no signs of irrigation having been practiced, showing that there was once sufficient rain to make the country habitable.

"There are reasons for believing that forests existed in certain regions, where today the tree and the vine could not secure a footing, for the hills are denuded of their soil. The land which the Greeks and the Romans found so profitable to develop, is now largely a waste; and it is difficult to appreciate, from what we see at present, what certain ancient writers tell us about the land; for example Cicero, who said that 'the country is so rich and so productive that in the fertility of its soil, and in the variety of its fruits, and in the vastness of its pasture lands, and in the multitude of all things which are matters of exploitation it is greatly superior to all other countries.' " (Manilian Law VI.)

A little to the west of this upper Mesopotamian kingdom of Amaru, Butler points out that a pronounced change of climate is necessary in order to explain the history of Syria where an area of 20,000 square miles which is now practically "desert and deserted—was more thickly populated than any area of similar dimensions in England or the United States is today if one excludes the immediate vicinity of the large modern cities. It has also been discovered that an enormous desert tract lying to the east of Palestine, stretching eastward and southward into the country which we know as Arabia, was also a densely populated country. How far these settled regions extended in antiquity is still unknown, but the most distant explorations in these directions have failed to reach the end of ruins and other signs of former occupation."

It would be easy to cite many other instances where a knowledge of climatic changes helps greatly in understanding history. In *World Power and Evolution* I have shown this in relation to Rome, while *The Character of Races* gives similar illustrations for China, Iceland, and other regions. But regardless of whether the historian feels the need of changes of climate to explain the historic record, no scientifically-minded historian can fail to take account of such changes if the specialists in such matters agree that they have taken place. The general attitude of such specialists, even in 1916, and still more at present, seems to be well summed up in the words of D. W. Johnson, professor of physiography in Columbia University. *"The Climatic Factor,"* he says in a review of that book, "does not solve all the problems presented by a theory of climatic pulsations. It does not pretend to do so. It does aim to show that the evidence thus far available strongly supports the belief that there have been within historic time climatic changes of a pulsatory nature. In this the author has, in the opinion of the reviewer, been successful."

One of the latest expressions of opinion is that of Antevs in *Post-glacial Climatic Cycles* (1924). "The view of distinct climatic fluctuations, advanced especially by Huntington in a number of papers, is being more and more accepted. The very best evidence of the occurrence of climatic fluctuations of short as well as long duration, up to thousands of years, is probably to be found in the varying rate of growth of the big trees and in the periodic retreat and halt of the edge of the Pleistocene ice-sheets. Huntington shows that there have been during historic times, within limited regions, climatic fluctuations of different length and character, but particularly in precipitation. The most noteworthy fact is that, especially in western and central Asia, the climate 2000 or 3000 years ago appears to have been distinctly moister than today."

CHAPTER XVI

THE SHIFTING CENTERS OF CIVILIZATION

THE final stage of our investigation is now before us. We have compared the present distribution of civilization and of climatic energy, and have seen their remarkable similarity. We have found that during historic times a pulsatory shifting of climatic zones appears to have taken place. Let us now examine the shiftings of civilization, and determine how far they are in harmony with those of climatic energy.

The black area of the map of energy, Figure 43, embraces about 1,100,000 square miles, or 2 per cent of the total land surface of the earth. Its population is about 260,000,000 or 17 per cent of the world's total. Consider what would have happened if all except this small 2 per cent of the lands had ceased to exist seventeen or eighteen centuries ago. Civilization would probably have been altered only slightly, although history would have been very different. If the rest of Europe had ceased to exist, Byzantine art, the works of the early Christian Fathers, Moorish architecture, and Spanish explorations, conquests, and colonization would be lacking. Yet this would not seriously change the conditions under which we now live.

Suppose that not only the more backward European countries, but all the rest of the world except the very high European area had been blotted out in 200 A.D. The influence of Greece and Rome would have persisted in Italy. The Roman Church, the Renaissance, and the Reformation would all have played their part. The absence of new lands for discovery in the Middle Ages would doubtless have retarded the development of certain ideas. Otherwise the status of civilization would have been changed only a little. The United States is the only non-European country which has succeeded in touching the daily life of the world as a whole. American inventions have gone everywhere. The denizen of Turkestan who discusses news received by telegraph and whose wife makes his clothes with an American sewing machine is being genuinely touched by the New World. Yet America must not boast too loudly, for without her aid, Britain, France, Germany, Italy, and their small neighbors would have made most of the world's great inventions. To an even greater degree they would have been able to develop the arts and sciences.

The energetic area of western and central Europe has been the great center of civilization for a thousand years. From it have gone forth more new ideas than from all the rest of the world combined. Its manufactures have flooded·all parts of the earth. Politically it dominates about 60 per cent of the earth's surface. Socially its domination is still greater. The United States has long been chagrined because South America has looked to Europe not only commercially but for her inspiration in literature, art, education, and almost every other phase of activity. Politically the United States helped greatly in inspiring the South American republics, and was at the front in stirring Japan to the new life of the last half-century. Yet those countries have turned their faces Europeward, because that is where ideas have been most numerous and have been most fully embodied in the concrete form of institutions, inventions, or scientific principles.

Wherever one turns, he feels the tentacles of the great European center of civilization reaching out and vivifying the life of the whole world. To tell why this is so would require a volume. The position of western Europe, the strong racial inheritance of its people, their legacy from the civilizations of the past, their inspiring religion, their political freedom, and their many powerful institutions, all play an essential part. With these, however, and as one of the conditions which make them possible, stands the fact that western and central Europe is the only large region which for many centuries has had these advantages and at the same time has enjoyed a highly stimulating climate.

The main American high area differs somewhat in shape on the maps of civilization and climate, but not essentially. In general, it occupies the portion of the United States lying north of about latitude 38° and east of the Rocky Mountains, together with the adjacent strip of southern Canada. Its area, taking the average of the black portions of the two maps, is approximately the same as that of the similar European region, that is, about 1,100,000 square miles. The population, however, is only 70 or 80 million. Here, then, we have another 2 per cent of the earth's land surface, containing between 4 and 5 per cent of the earth's total population. Consider how disproportionately great an influence it exercises. To be sure, it falls far behind Europe, but that is partly because it is newer and less densely populated. Yet it differs from Europe only in the degree, not the kind of influence. We have already referred to some of its inventions. All the world recognizes that the telephone and telegraph and many other forms of applying electrical energy are primarily American. In architecture we have evolved such wholly new types as the sky scraper. In art and literature we follow far in the wake of Europe. Yet we have done some things which are widely known in other countries, and which are steadily molding the world's opinion. The same is true in science, for the day is past when Europeans can ignore the work done on the western side of the Atlantic. In the future the disparity between Europe and America along this line bids fair to decrease rapidly, for great institutions devoted to pure science are growing on a scale unknown in other lands.

This is in part the result of another line of effort in which Americans have achieved unusual success. By this I mean what is sometimes called "big business," the amalgamation of many minor industries or branches of an industry into one huge whole. It is illustrated in such organizations as our great railroads, and the United States Steel and Standard Oil companies. These represent in part the effect of unusual opportunities, but they also indicate a great expenditure of energy in planning and directing such widely ramifying activities. Lastly, and perhaps most important of all, the United States has touched the world upon the political and social side. In spite of political sins, which are many, this country has stood for freedom and justice. Its example has inspired other nations, including parts of western Europe, and has given them a stronger desire for equality of opportunity and for the breaking down of unfair discriminations between man and man. In more backward places a similar influence is felt. Good judges say that in Turkey, for example, one of the most potent causes of the revolution of 1908 and the establishment of representative government was the teaching of American missionaries. The same is true in China, except that there the Americans have been associated with missionaries of many other nations, while in Turkey they have been almost alone. In many other ways the ideas of this small area radiate in all directions. In that lies the essential point. Because this 4 or 5 per cent of the world's population are endowed with unusual energy, they are rapidly obtaining an influence wholly disproportionate to their numbers.

The third important region of high civilization and energy is Japan. Although the area is only about one tenth that of the corresponding portion of Europe and America, the population is over 50,000,000. In proportion to population, its influence upon the world is less than that of the European countries and the United States, which perhaps corresponds with an apparently lower degree of climatic energy. Yet we all recognize that Japan is a factor to be reckoned with. We admire its art and the skill and certainty with which the country has made use of our western civilization. We see students flocking thither from China. In India the news of the progress of Japan causes a stirring of dead leaves as by an autumn wind. As Japan has done, so India would do. In India, however, the agitators chiefly talk and commit violence, whereas in Japan they go to work quietly without much talk or violence and make themselves so efficient that they do not need the help of Europeans. When Japan sees an opportunity, such as was afforded by the war in 1914, she takes it, and thus advances another step in her rôle as the most capable nation in Asia. Like Europe and America she makes mistakes sometimes, but they are the mistakes of strength more than of weakness.

Next to Japan among our areas of high civilization comes New Zealand, together with the adjacent corner of Australia. Here we have chiefly a reflection of British civilization. Most of the five or six million people are grouped in an area somewhat more than twice the size of Japan. The question is not whether they show evidences of a high civilization, which they certainly do, but whether they have added to the endowment of ideas and institutions with which they came from England. They have clearly done so, for when we think of Australia there come to mind such things as the Australian ballot, whereby a man is enabled to vote freely without undue influence from bystanders. We think also of progressive labor legislation, and of a pioneer attempt at old age pensions, especially in New Zealand. We are made conscious that these people are not mere followers, but are blessed with ideas which they themselves originate, and which spread out from them to other parts of the world.

The last area which rises to the very high grade on the map of civilization is California and the coast farther north. The population is still relatively small, and the majority grew up in other regions. Hence, it is too soon to judge of the ultimate status of this area. It is so bound up with the rest of the United

States and Canada that its particular contribution to civilization cannot be clearly differentiated. Yet, through their universities and their readiness to attempt great things, the people of the Pacific coast impress the world as unusually capable and energetic. All together these five regions comprise less than one twentieth of the lands. Though they are densely populated because of the skill of their people in agriculture and manufacture, they contain only about one fourth of the world's population. In influence, however, they many times outweigh the remaining three fourths. If they decide on anything the rest of the world must submit. They never care to imitate the more backward nations, whereas such nations show a growing desire to imitate Europe, the United States, or Japan. The reason seems to be that the favored one twentieth of the land is inhabited by people of great energy; and this energy appears to be due, in part at least, to the frequency of cyclonic storms with their stimulating changes of weather from day to day.

This leads to the question whether similar conditions prevailed in the past. Throughout the course of history there have been centers of civilization like those which exist today. One of the most important was the plain of Mesopotamia. Here, in the earliest days, the people of Sumer and Accad invented hieroglyphs, founded great cities, built mighty irrigation works, drained the marshes, and laid the foundations of a complex system of law and religion. They clearly possessed the power of originating ideas and of putting them into effect in a way that is now characteristic of the modern centers of civilization. Where they originally came from we do not know, but apparently they were not Semitic. In course of time, they were overwhelmed by Semites who presumably came in as rude invaders. Then there happened one of the most characteristic events in the history of centers of civilization. The Semites dropped their old culture and adopted that of the people whom they had conquered. They began also to be inventive like their predecessors, and thus de-

veloped cuneiform writing out of the hieroglyphics of the Su-
merians. In later days, other Semitic invasions took place. In
each case the invaders had done no great things in their original
homes so far as we know, but when they came to Babylonia they
blossomed into people of inventive minds full of new ideas which
caused the arts to rise to higher and higher levels. Then the
Semites were conquered by a non-Semitic race, the Scythians,
who were also called Medes, and they in turn were stirred to great
achievements. They were conquered by still another people, the
Persians under Cyrus, and once more the old process of stimulation was repeated.

One of the most significant things in Mesopotamian history
is the way in which each invading race seems suddenly to have
risen in civilization as soon as it reached the new country. How
much of this was due to natural selection, whereby the weaker
people among both the conquerors and the conquered were
weeded out during the migrations, wars, and pestilences which
preceded the final settlement; how much to contact with a
higher civilization; how much to more abundant opportunities
afforded by the new environment; and how much to a direct
physical invigoration? I shall not try to answer these questions.
That some of the invading races achieved great things because
they possessed innate ability seems certain. That their minds
were rendered more alert by contact with the achievements of
the races which they conquered is likewise beyond doubt. That
the wealth and agricultural possibilities of Mesopotamia and the
abundance of labor stimulated the invaders to construct palaces
and temples, build irrigation works, gather stores of treasure,
carve huge monuments, spend money in supporting libraries
and students, and carry on other great enterprises also seems
certain. These principles are so firmly established that there is
no need to discuss them. The question before us is simply
whether there may also have been a climatic stimulus compa-

rable to that which now seems so important according to our maps of civilization and energy.

I do not purpose to discuss the empires of antiquity. I merely wish to point out some of the features of their location and history which pertain to our main hypothesis. While Babylonia was flourishing, Egypt in the same latitude was also developing a great civilization. The two were rivals. Which rose higher it would be hard to say. They moved along parallel lines, and at last both fell during the same general period, although Egypt retained a moderate importance for some centuries after Mesopotamia had lapsed into insignificance. Far to the south the Sabæans in southwestern Arabia also developed a fairly high civilization, but they formed in no sense a great center. We know little about them, but apparently they were behind Mesopotamia and Egypt in the way that modern Bulgaria is behind England. Therefore, we may pass them by for the present although they will come up again in another connection. East of Mesopotamia the highlands of Persia were the seat of a comparatively high culture, but it scarcely rivaled that of the two great countries in the river valleys. Going on still farther toward the rising sun, we find high civilization in northern India, at a very early time. That region never rose to such a pitch as did the countries farther west, and it decayed somewhat earlier. Beyond India the chain of ancient empires is broken until we come to China. In Indo-China, to be sure, traces of a state of culture higher than that of today are found, but they belong to a comparatively late date, and it is doubtful whether they were indigenous. They are important, but do not apply to the period now under discussion. China, however, stands in the same class with Egypt and Mesopotamia. It reached a high state of development about as long ago as they did. Here, too, the art of writing developed, inventions of many kinds were made, strong governments were organized, other powerful institutions were evolved, and in general there was intense mental activity.

At this point the reader is advised to consult a map of Asia and Europe. Notice how the great countries of the past lie along an arc. They begin on the east with China, whose main center was north of the Yangtse basin. The civilization which flourished there possessed qualities like those of modern Europe and America. The Chinese were an aggressive, active people, able to fight or to evolve a new religious system as the case might be. The sword was in their hands not merely because those were ancient days, but because the people or their rulers wanted to make conquests, or because they were quick to resent infringements of their rights. They were not like their modern successors who see themselves defrauded because they have not the power to cope with their despoilers. Follow the map westward along a band eight or ten degrees wide with the thirtieth parallel a little south of its center. At first we find ourselves among lofty mountains which merge into the great Tibetan plateau. Here we should not expect any great development because of the unfavorable topography. At length, however, our band descends to the plain of northern India. On the flank of the Himalayas lived Gautama, the great thinker who founded Buddhism. Out on the plains around Delhi and Agra and in surrounding regions was the home of the early Aryans, who developed the Sanscrit language, wrote the Vedas, and made a real contribution to human progress. Here, just as in China, a broad expanse of arable land was favorable to the development of a great people. Next our band strikes the high mountain rim of Baluchistan and Afghanistan, and then passes over the plateaus of those countries, high and rugged in the north, lower and intensely desert in the south. In these regions the ancient civilization was higher than that of today, but not until we reach the western side of the plateau do we come upon the really advanced country of ancient Persia.

As our band leaves the Persian highland, let us change its direction a trifle so that its central line will point toward the center of the Mediterranean Sea. First it lies over Mesopotamia. There in the highly fruitful plain we find the two great empires of Babylonia and Assyria. They were like their modern successors in the Europe of our own day. When they were weak, they were invaded by the more backward people from the mountains on the north and east or from the desert on the south and west. When they were strong, they waxed ambitious; they strove to extend their borders; they desired to rule the world; and they were jealous of rival nations. Thus, at their strongest, like the great European nations which were locked in a death struggle only a few years ago, they fought tooth and nail for supremacy. The fighting in itself is a proof of superabundant energy. A weak people may fight in self-defense, or to gain new lands when sorely pinched with want at home, but only a people who have a surplus of both strength and wealth can stand the strain of great wars whose purpose is primarily national aggrandizement.

On the flanks of our band of culture, as we call it, lay the Hittites on the borders of Mesopotamia and Anatolia. Beyond the Syrian desert the Phœnicians, Syrians, and Jews lived in the center of the band. Their degree of culture was equal to that of the Mesopotamians, but displayed itself in different ways, partly because of the accidents of mountain chains and seas, and partly because of peculiar circumstances of their racial character or history. Next comes Egypt, still within the belt. Here still another fertile plain gave opportunity for the accumulation of great wealth and for the leisure which is essential to a part of the population if new ideas are to be developed. They fought the same kind of wars as did their neighbors on the east. When they were most powerful, their strength and energy were such that the mere conquest of their immediate surroundings did not satisfy them. Taking Palestine, they penetrated Syria and even advanced to the Euphrates to fight with Assyria, their great rival. And the Assyrians in turn came down through Palestine to wrest the sovereignty from Egypt.

Now we pass on, curving our band a little more, so that its center shall pass along the main axis of the Mediterranean Sea. Rhodes, the Ægean Sea, Greece, Italy, and Carthage fall within its limits. Here, too, we see the same phenomena of the rise and fall of great nations. The rise did not begin quite so early in the west as in the east, and the fall did not come so soon. In Greece, as in Mesopotamia, a growth in strength brought with it a great struggle for supremacy, which did not end till Athens had subdued her rivals. The wars of Rome and Carthage were of this same kind in which great nations, jealous of the greatness of a rival, well-nigh destroy themselves.

In this rapid sketch we have mentioned each of the chief civilizations of Asia and Europe up to the time of Christ. All were in their day centers comparable to those which we have seen to exist at present. One main center lay in China, and corresponded to that of modern Japan. A second, less important and less enduring, was located in northern India. It has no modern representative, unless we hold that the vogue of an esoteric Hinduism among a few people in America and Europe is a sign of the existence of a widely pervasive influence derived from an Indian center. The third center extended widely. Its eastern limit was on the western border of Persia. It embraced Mesopotamia, Syria, Egypt, and Greece, and on its flanks included surrounding countries such as Asia Minor. Ultimately it moved westward, so that Mesopotamia and then Egypt ceased to belong to it, while Italy was included. Finally, it migrated still farther westward and also northward until now Italy is on its southeastern border, and its main area is the great European center of our own time.

In the western hemisphere only two centers of civilization seem to have existed in ancient times. One was in Peru. We shall not consider it here, partly because it will come up in another connection, and partly because it failed to rise to a great height, and never succeeded in spreading its ideas very widely. The other

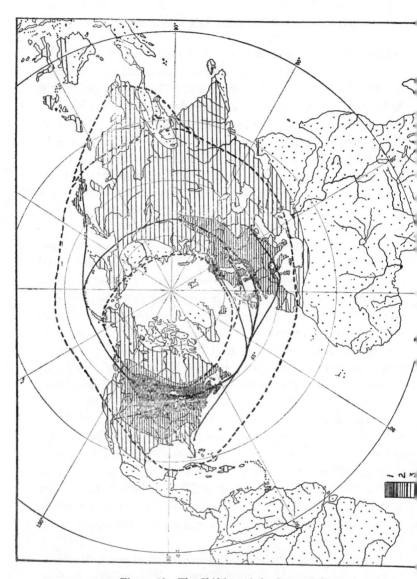

Figure 48. The Shifting of the Storm Belt

1. Regions of Great Storminess (Over 20 Centers per Year According to Kullmer's Scale).
2. Regions of Moderate Storminess (10-20 Centers).
3. Regions of Slight Storminess (1-10 Centers).
4. Regions Without Cyclonic Storms.
The Median Lines of the Main Storm Area and its Branches are Indicated by Heavy Solid Lines.
The Hypothetical Median Lines of the Ancient Storm Belts are Shown by Dotted Lines.

ancient American civilization, that of the Mayas, is much more important. It falls into the same class with those of Eurasia. Its fertility in ideas has already been discussed. Its influence spread to other lands, and may be seen in various ruins in Mexico, and in the calendars which were employed by several races in that country. It must have lasted many centuries, for we find splendid buildings whose dates are at least seven hundred years apart and possibly more. Moreover, before the first of these was constructed there must have been a long period of development during which the arts of architecture and sculpture were perfected. At that same period the Mayas must have been so far advanced that they kept most accurate records of astronomical phenomena, for otherwise they could not have determined the length of the year with the extraordinary accuracy displayed in their calendar. In all these respects the Maya center resembled those of Eurasia. It differed from most of them in not leaving a successor.

So much for the ancient distribution of civilization. Now for that of climatic energy. Storms, as we have seen are the most variable of the climatic elements and also one of the most important in their effect on human energy. Figure 48 shows the approximate distribution of storminess in the northern hemisphere at the present time. It is constructed with the north pole in the center to show the complete storm belt encircling the north temperate zone. The lands are shaded to represent three grades of storminess. Heavy lines indicate the centers of the main storm belt and its branches. The region of maximum storminess lies in southern Canada north of Lake Ontario. Thence the main belt extends eastward along the St. Lawrence River to the southern edge of Newfoundland, where it is joined by a branch coming up the Atlantic coast. Then it crosses the Atlantic with a slight inclination toward the north, and before reaching Ireland splits into three branches. One passes along the northern edge of Scotland, through northern Scandinavia and thence along the Arctic coast of Europe toward Asia where it almost dies out and probably merges with the second or main branch. The latter splits into two parts which unite again. One passes through Scotland and southern Scandinavia. The other goes through southern England, follows the Baltic Sea and then, after joining its companion, extends eastward in the latitude of Moscow for an indefinite distance into Siberia. It apparently passes entirely across Asia, but in an attenuated form. The third branch swings from the central Atlantic toward the Bay of Biscay. It becomes weakened over southern France, strengthens once more in the stormy Gulf of Lions, continues with great intensity in northern Italy, then loses intensity as it crosses the Bosnian highland, and finally passes through southern Russia where it rapidly dies out. Returning to the main belt of storminess we find that after it has crossed Asia it is joined in the Pacific Ocean by a branch which has been highly developed in Japan, but which weakens northward. The combined branches cross the Pacific Ocean to the American shore. Then in southern Canada and the northern United States, after crossing to the east side of the Rocky Mountains, the belt once more becomes extremely stormy. It is joined by a branch from Colorado and so completes its course back to the point of greatest intensity.

Several features of Figure 48 deserve careful attention. In the first place, wherever the storms in their eastward course approach an extensive highland they tend to be weakened or deflected, after which they may or may not recover their former intensity. This may be seen on the western side of North America where storms are relatively rare until the Rocky Mountains have been crossed. The highlands of western Europe are not great enough to destroy the storms, but the main belt is broken into several branches, each of which seeks to avoid the high country. The northern branch would apparently skirt the whole of Scandinavia were it not that some unknown force prevents it from going much to the north of the Arctic circle. The next avoids the Welsh and English highlands, crosses into the North Sea over the lowest part of England, receives an accession of intensity, and follows the line of least resistance along the Baltic and across the plains of Russia. Ultimately, however, the vast expanse of Eurasia, especially when it becomes a high pressure center in winter, largely overcomes the tendency of the air to move in the cyclonic fashion, and storms become rare. The third branch, which swings still farther south, is also greatly weakened when it reaches the land of southern France, but becomes very pronounced in the northern Mediterranean, only to be weakened once more by the Balkan highland.

Another important feature of Figure 48 is the peculiar detached areas of high storminess in northern Italy and Japan. In the United States two similar areas are visible on a large scale map of storms, although they are masked in Figure 48. In both cases they are apparently remnants of a southern, or subtropical belt whose existence at times of many sunspots is described in *Climatic Changes.* Another belt where storminess increases greatly at times of sunspot maxima is found on the north side of the present belt. It tends to reduce the habitability of the northern parts of Europe and North America, but with that we are not now concerned. In the southern belt of increased storminess, when sunspots are numerous, the American areas move somewhat southward, become intensified, and lengthen east and west until they coalesce, forming a continuous, although very faint, belt. In Eurasia the same thing seems to happen, although the meteorological records in Asia are not yet sufficient to permit us to test it. If the American subtropical belt were intensified in the way that appears to have occurred in the past, its place of maximum storminess would probably be in the Gulf of Mexico. There the obstacles presented by the highlands of the southwestern United States and Mexico give place to the open sea and allow free play for the development of storms. A development of this subtropical storm belt seems to be a pronounced feature of the changes which now take place from sunspot minima to maxima, and would presumably be magnified if the changes were on a greater scale than those of the 11-year cycle. Hence, we should infer that when the Mayas made their greatest progress, their country was blessed with a stormy area like that of the Gulf of Lions which now does much to make northern Italy far more progressive than the southern part of the country. Under such conditions the climate of Yucatan would have been comparatively stimulating in spite of its warmth, for during much of the year cool waves from the north would have been frequent.

In Eurasia the same principles apply with appropriate modifications because of the topography and size of the continents. If the subtropical storm belt should be magnified in the way that seems characteristic of times of many sunspots, it would apparently reach its chief development in the places where the great civilizations of the past were located. The present Italian center would presumably move south and then expand eastward along the open Mediterranean. Leaving the highlands of Asia Minor on its northern flank, it would pass over the Syrian mountains because it could not spread eastward in any other way. Possibly it would be forced a little southward so that northern Egypt and Palestine would get its full force. In the lee of the mountains the storminess of the Syrian desert would be less than that of Syria, but by the time the storms reached Mesopotamia they would have had an opportunity to form themselves anew, and we should expect a well-developed center. In Persia, judging by what happens today, this would be weakened but not destroyed. Then the great plains of northern India would permit the storms to gather once more in a center corresponding to the faint one of today. That would probably be the end of any distinct belt for a thousand miles, for the eastern Himalayas and the mountains of western China interpose so tremendous a barrier that we should not expect the Japanese center, even though it moved southward and westward, to coalesce with the Italian. The plains of central China, especially the Yangtse valley, would be the place where the present Japanese center would be most highly developed, provided it suffered changes corresponding to those indicated in the corresponding American area. Japan, on the other hand, would presumably be less stormy than now, for the storm tracks would be pushed southward and oceanward.

If these things actually occurred at the times indicated by our California curve, the countries which now stand highest in civilization would then have had a long, cold winter, ending in a very stormy spring and followed by a cool damp summer. The stimulating qualities of the climate would have been less than now, and the possibilities of agriculture, especially in such places as Germany, would have been much diminished. In the countries like Greece where civilization was then at its highest, on the contrary, the number of storms and the duration of the stormy period would have been decidedly greater than now. The rainfall also would have been greater so that agriculture would have been favored. More important than this, however, would have been the high degree of climatic stimulus because of frequent changes of temperature. To apply the matter concretely, lower Egypt has an average temperature of about 80° during July, and Mesopotamia and northern India are even hotter. Three thousand years ago the heat was probably almost equally intense at certain periods, although somewhat less extreme on an average, and by no means so uniform. All the summers were probably hotter than was the summer of 1911 in the northeastern United States. That summer, in some localities, was the most severe for a century. Yet its effect on work was no worse than that of an ordinary winter. In Mesopotamia and Egypt, summers of that kind doubtless made people slow and inert. They were probably not so bad as the present summers, however, for although no appreciable quantity of rain may have fallen, a large number of storms must have passed over regions not far to the north. Such storms would stir the air and bring fresh breezes. Everyone knows that in hot weather a change in the wind even without rain is most refreshing. When the summer was over, the storm belt, according to our hypothesis, would migrate southward in its normal fashion, and soon Egypt and Mesopotamia would be swept by storm after storm. During the early fall and late spring conditions would be about as they are in the homes of our Connecticut factory operatives during cool summers such as 1912 or 1913. For the intervening five or six months the average temperature would range from about 55° to 65° F., there would be a constant succession of cool waves, and the conditions would be almost ideal for great physical activity. Thus, even though the summers were distinctly bad, the total debilitating effect would be little greater than that of summer and winter combined in Connecticut. In Greece and Italy, with their more favorable mean temperature, conditions would be still better than in Egypt or Mesopotamia. In the same way favorable conditions appear to have prevailed in each of the great countries of the past at the time when it made its most rapid progress.

The two phases of our climatic hypothesis are now before us. In point of time, though not of presentation in this book, the first step was a study of the climate of the past. Ten years of work along this line had led to the hypothesis of pulsatory changes, and finally to the idea that the changes consist primarily of a shifting of the belt of storms. After this conclusion had been reached, a wholly independent investigation of the effect of present climatic conditions upon human activity led to two conclusions, neither of which was anticipated. One was that in the eastern United States the cold weather of winter is as bad for work as for health, while only the warmest summers cause any serious curtailment of work. The other was that storminess and variability from day to day are of great importance. On the basis of these two conclusions it at once becomes evident that the stimulating effect of climates in the same latitude and having the same kind of seasonal changes may be very different. It also becomes clear that the distribution of civilization at the present time closely resembles that of climatic energy. From this the next step is naturally back to our previous conclusion that changes of climate in the past have consisted largely of variations in the location of the storm belt. If this is so, evidently the amount of climatic stimulus must have varied correspondingly. Thus we are led to the final conclusion that, not only at present, but also in the past, no nation has risen to the highest grade of civilization except in regions where the climatic stimulus is great. This statement sums up our entire hypothesis. It seems to be the inevitable result of the facts that are before us. Other factors, to be sure, are also highly important. One of them is natural selection, a subject so much neglected in its bearing on human history, that I made it the theme of a book on *The Character of Races*, which might almost be considered the continuation of the present book. The development of human culture and its spread from land to land is, I believe, a third factor quite as important as either of the others. Yet, unless we have gone wholly astray, the surprising way in which independent lines of investigation dovetail into one another seems to indicate that a favorable climate is one of the essential conditions of high civilization.

CHAPTER XVII

ABORIGINAL AMERICA AND MODERN AUSTRALIA

HAVING seen that the objections to climatic pulsations are largely based upon misapprehensions, let us now take up certain other objections to our hypothesis of the relation between civilization and climate. "How about the high civilizations of the past in places outside the storm belt?" the objector may say. "Yucatan and Guatemala may possibly have been stormy, but surely not Peru, southern Arabia, Rhodesia, Ceylon, Java, and Indo-China. Yet in the past these have boasted a civilization much higher than now prevails there."

The force of this objection must frankly be admitted. The ruins of Indo-Chino and northern Ceylon, and any other traces of high civilization located in tropical lowlands, with the exception of Yucatan and Guatemala, appear to have little relation to our hypothesis. Nevertheless, they do not require a modification of it. More probably they represent the triumph of other factors over the climatic factor. They apparently indicate one or both of two things. First, they appear to represent a temporary wave of progress due to the incursion of higher cultures from more favored regions such as India or the neighboring highlands. If that is so, they possess no more significance for our climatic hypothesis than do the railroads and other innovations introduced by Europeans within the past few decades. In addition to this they probably indicate that prolonged migration or other difficult circumstances gave rise to pronounced natural selection. Thus the people who built the wonderful ruins of Java and Indo-China, as I have shown in *The Character of Races*, and perhaps those of Ceylon, Peru, southern Arabia, and Rhodesia, had presumably been gradually sifted so that only a very competent remnant was left. Under such circumstances the effects of climate seem to be overcome for a while, but reassert themselves in due time as the strong inheritance becomes diluted or weakened.

Another possibility is that the small and insignificant belt of storms which now traverses northern India may have been intensified in the past. In eastern India because of the barrier of great mountains, a branch may have swung southeast around the lofty Burmese mountains to the neck of the Malay Peninsula and then northeast to join the storm belt which now traverses Japan. In crossing the plains of Cambodia, where the most remarkable ruins of southeastern Asia are located, it may have developed a distinct storm center like that of northern Italy. The conditions would resemble those inferred in Yucatan and Guatemala, although perhaps less favorable because cold waves would be checked by the great mountains. Indo-China and Central America lie in the same relative position in their respective continents, a fact which may have considerable significance. I do not, however, attach much importance to what has just been said about storms in southeastern Asia, for it is pure supposition. Other conditions, such as racial selection and the spread of the highly developed culture of India seem competent to explain a large part of the sudden rise of these regions, while the tropical climate largely explains their rapid decline.

Yemen, Rhodesia, and Peru furnish admirable examples of comparatively high civilizations which developed in tropical highlands and lasted for a considerable period without much intercourse with regions of higher culture. They lie at such an altitude that the mean temperature never departs far from the optimum. For instance, at Fontein, near the Rhodesian ruins of Zimbabwe, the figures are 54° F. in June and 71° in November. For Yemen none are available, but judging by the altitude its temperature is probably slightly lower than that of Rhodesia.

Even at Cuzco the elevated capital of old Peru, although the monthly mean temperature according to Hann ranges from 46° in July to 52° in November, the conditions are very stimulating compared with tropical lowlands. The temperature is too low for comfort, to be sure, but it is of the type that appears to stimulate mental activity. If immigrants possessed of an unusually high inheritance, either by reason of strenuous natural selection or of actual biological mutations, should come to such a region, the relatively stimulating quality of the climate would combine with their innate ability to enable them speedily to dominate the indigenous population, and to develop many new ideas. Yet we should not expect such a civilization to endure so long or rise so high as those in more favored regions. The people would tend to exhaust themselves for they would never experience any restful changes of seasons, and would be stimulated at all times. To revert to an earlier illustration, their condition would be like that of a horse which is always driven at full speed. Such a horse might go rapidly for a while, but would wear himself out at an age when carefully driven animals were in their prime. The constant nervous excitation produced by such a climate in immigrants from a less favorable region would induce both progress and decay. The most nervous people would die out partly because they would exert themselves too strenuously, and partly because nervousness is a potent agency in reducing the birthrate. Self-control would also be weakened, thus leading to vice and excesses of various kinds. The chewing of the coca leaf, a narcotic stimulant which is thought by many people to be one great cause of the backwardness of Peru today, would harm the high-strung, competent parts of the community more than the dull, apathetic ones. In various other ways such a climate as that of Peru would be stimulating for a while, but would lead to exhaustion sooner than would one where greater variability prevails.

We now come to perhaps the strongest objection aside from

the question of the reality of climatic changes. A large part of this book has been concerned with the United States. The present status of that country is one of the main foundations which originally led to the framing of our hypothesis. Yet before the coming of the white man savagery prevailed where civilization is now highest. In the sixteenth century the climate must have been approximately the same as that of today, so that no explanation can be sought along that line. It is true, to be sure, that before the coming of the white man the most energetic and capable of the Indian tribes lived in what is now the northeastern United States. Dixon, for example, in *The Racial History of Man*, says that the "Iroquois and southern Algonkian tribes were among the first of those north of Mexico in ability and prowess." Many other authorities speak in similar fashion. Wissler in *Man and Culture* reproduces a map by Moorhead showing that the region from Illinois to New York was an important center of culture from which many new types of stone ornaments spread in all directions.

In spite of this we are still confronted by the stubborn fact that before the coming of the white man no great civilization had ever developed in the northern United States, while Mexico, although far inferior climatically, harbored a relatively civilized population. At first this seems to prove that our hypothesis cannot be correct. Yet more careful examination leads to a different conclusion. The distribution of civilization in pre-Columbian America merely brings out a fact which I have again and again tried to emphasize: although climate is highly important, there are other factors whose weight is equally great. Even if our hypothesis be fully accepted, no less importance will thereby attach to the other great factors which condition the events of history. Because a man dies for lack of air, we do not think that air is more essential than food, drink, warmth, the circulation of the blood, the reproduction of the species, the ravages of virulent germs, and many other conditions. So it is in history. Even if our climatic ideas are correct, it will still be true that the ordinary events of the historical record are due to the differing traits of races, the force of economic pressure, the ambition of kings, the intrigues of statesmen, the zeal of religion, the jealousy of races, the rise of men of genius, the evolution of new political or social institutions, and other similar circumstances. Yet a comprehension of the part played by the climatic factor will enable us to explain some of the many events which hitherto have puzzled us. Not all will be thus explained, for others must wait until the action of some other set of as yet unknown conditions is understood.

In the present case I believe that the explanation of the failure of the Indians to rise to a level equal to that of the people of the Old World is found partly in inheritance, and partly in the history of their culture. I shall not dwell on inheritance, although it may be of great importance. In *The Red Man's Continent* and more fully in *The Character of Races* I have set forth the hypothesis that the aboriginal Americans during their migration from Asia by way of Siberia and Alaska suffered an adverse natural selection due to the severe arctic environment. This may have given them a permanent handicap compared with the people of Europe and western Asia among whom the selection was peculiarly favorable. Among the Americans, so it seems, there was a premium on powers of passive resistance and endurance. Moreover, among all hunting tribes there is a premium upon the ability to make sudden but not prolonged efforts such as are involved in hunting and raiding. Only when agriculture is introduced does a new type of natural selection put a premium upon such qualities as steady industry. Among hunting tribes, moreover, the acquisitive temperament which stores up food or other property for periods of want, is not nearly so great an advantage and hence so strong a selective factor as among those who practice agriculture. Because the wandering hunter cannot transport much property, he loses the biological advantage which causes the people who have a careful, thrifty spirit to have a greater chance of survival than have the careless and unthrifty.

Turning now to the cultural side of early America, at least two great factors appear to have imposed a tremendous handicap upon the Indians of the northeastern United States. One is the absence of tools of iron, and the other the lack of beasts of burden. These, in conjunction with the type of vegetation, offer such hindrance that the Indians could probably never have developed a very high civilization, even if they had been as competent as the races of Europe. The discovery of the use of iron depends on the coincidence of three conditions whose occurrence is almost accidental. One is the birth of a man with sufficient inventive genius to devise a way of obtaining the metal from rocky ores. The second is the occurrence of ores and other necessary materials within easy reach of such a man. The third is that the genius must be so relieved from the fear of enemies, the danger of starvation, and the ravages of disease that he has both time and strength to elaborate his idea. The combination of these three fortunate circumstances never occurred in America. Although a little native copper was used, we have no assurance that the ore was ever smelted in any large amounts. It certainly never was used to any great extent. Iron, except a few bits from meteorites, was absolutely unknown. Hence, through no fault of their own, the original Americans were dependent upon such tools as they could fashion from flint, obsidian, and other stones, or from bones, shells, wood, and similar materials.

We are so accustomed to iron tools that we scarcely realize their immense importance. Consider their effect upon agriculture. Go into a virgin forest with its labyrinth of trees. Imagine the task of cutting them down with a stone hatchet. The mere physical labor is such that none but people of high energy would ever attempt it. An easier way is to girdle the trees, cutting off the bark in a circle, and then leaving them to dry until they can be burned. Even that, however, is a long process. In the moister parts of the world its difficulty is greatly increased by the fact that while the trees are becoming dry enough to burn, new vegetation is rapidly growing. In many tropical regions, as we have already seen, the clearing of the forests is scarcely feasible even today. In temperate forests the difficulties are not so great, but they are practically insurmountable from the standpoint of a savage who has never known the meaning of hard, steady work. In the hunting stage a man may follow the trail until he falls from exhaustion, but that does not give him the power to wield a stone axe day after day upon stubborn trees. Hence, if agriculture is to be practiced by primitive people, it must originate in regions where there are no forests. The vegetation must be such that the primitive savage with his hands or with a stick or stone can easily grub it up. Or else it must be so scanty that it will not interfere with his crops. Anyone who has seen the agriculture practiced by people who are just emerging from some other mode of making a living knows that the fields are sown in the spring and then generally left untended until the time of harvest. Since the people are not yet able to get their living entirely from the crops, which is always the case during the transition period, they must perforce carry on the old occupations while the crops are growing. Hence, the weeds are allowed to grow as freely as the grain. In a dry climate where the natural growth consists only of bunchy grasses or low bushy weeds which grow apart from one another and can easily be pulled up with the hands before the next sowing, this does little harm. In wetter regions, such as the forested areas, the weeds are much more difficult to eradicate.

The reader has probably noticed that I have said nothing about grass-lands. The reason is that primitive people never attempt to cultivate them. If grass covers the entire ground and forms what we commonly call sod, crops cannot be raised in it. The dry grass may be burned off in the early spring, and seeds may be dropped in holes punched with a sharp stick, but no crop will be reaped. The roots of the grass are not killed. The new blades shoot up at once, and by the time the seeds begin to sprout are so high as to strangle them. The primitive savage who has no iron tools cannot possibly dig up the sod. Even if he has such tools, the laboriousness of the task bans it as a practical method of making a living. If anyone doubts this, let him spade up a turfy spot ten feet in diameter. Then let him calculate how many days he would have to dig in order to make a living for himself and his family by sowing grain. Let him also determine how a primitive savage who had to do this task would support himself and his children while he was digging before he could get a first crop. Finally, remember that the savage had no good spade to help him, but only a clumsy, heavy stone set in a handle, a brittle bone, or a piece of wood rudely shaped into a flat implement.

I have emphasized the difficulties of grass-land because they apply directly to the problem of civilization among the Indians. The part of North America which is highly stimulating climatically is covered either with grass of the prairie and great plains type, or with temperate forest. It is universally recognized that a high state of civilization is impossible unless it is based on agriculture. Otherwise people must wander all the time. They cannot accumulate the appliances which are essential even to the lowest real civilization. They are not attached to the soil, and thus have no incentive to improve a particular tract. If they possess domestic animals, the case is better, for these foster the sense of ownership and various other feelings which serve as uplifting influences. In North America no domestic animals except the dog and the turkey were known to the aborigines. This was not the fault of the Indians. The horse and camel families were then extinct in North America. The bison was the only animal of the ox family with which the Indians came in contact. It was too large and fierce to be domesticated.

The mountain sheep might possibly have been tamed, but it is very wild, its habitat is the barren mountains, and it is too small to plough sod. Even if the Indians had been familiar with the idea of taming beasts of burden, they could not have achieved any important success.

Without beasts of burden ploughing is impossible. Unless the ground is ploughed, grass-lands, as we have seen, cannot be cultivated. Therefore, civilization could not flourish in the grassy plains of North America before the coming of the white man. A few highly favored spots such as river flood plains where the grass grows in bunches and does not form turf were cultivated, but the areas of that kind were too small and scattered to give rise to any widespread progress. They were exposed to the ravages of Indians who had not yet learned the value of permanent abodes and individual property, two primary requisites of civilization. The people who lived upon them were swamped in the flood of surrounding savagery. Moreover, the places which they were able to cultivate were subject to floods and other disasters, so that the cultivators were often forced to rely on the chase. Then again, without iron tools or beasts of burden, a man cannot cultivate a large area because the task is so great. Therefore, the crops were likely to be insufficient to support the family throughout the year even when the harvest was good. The only recourse was the chase. In Asia people who were in such circumstances might rely on domestic animals, but not in America. Hence, it was practically impossible for the Indians of the grassy plains to get away from the savage life of the hunter.

In the forests which covered all the eastern states conditions were scarcely better. A man might girdle enough trees and cut enough brush to make a small field. When the wood was dry, he might burn it and obtain a crop of corn. While it was maturing he and his family would make a living by hunting and fishing. The crop would support them through the winter, but it is extremely doubtful whether they would ever cease to regard hunting as a main source of livelihood. The reason is this. The first crop on a burned piece of forest land in the eastern United States meets with little difficulty. For the second crop it is necessary to cut down and burn the weeds and bushes that have grown up during the preceding summer. Remember that the savage has no iron hoe, his seeds are planted haphazard here and there among the stumps, and he cannot tend his garden while it is growing, but must go off and hunt. When he cuts the bushes and weeds the second spring he does not root up the grass. He has no spade with which to dig it up, and no ox or donkey to draw a plough. If he takes good care of his field he soon finds that he has a meadow of tough sod on his hands. If he neglects it, he has a few acres of bushes. In either case if he would make a living from agriculture, he must go through the labor of clearing a new field. Meanwhile he still must be a nomadic hunter. He takes no great pride in his field; he does not care to improve it; he rarely builds a permanent house beside it. Why should he? Next year, or at most in two or three years, he must build another. In a few specially favored spots the conditions may be such that agriculture is not quite so difficult. In the North, where the stimulating climate is found, such spots are few and small. The chief reason for wonder is that the Indians had energy to cultivate any crops whatever. In the southern states the grass does not form so dense a sod as in the North, but even there the difficulties are great. Yet a beginning had been made, and the civilized nations, including the Cherokees and others, lived in permanent villages and practiced agriculture as their main means of getting a living. If they had had iron tools and oxen, America today might be filled with highly civilized Indians instead of Anglo-Saxons and other Europeans. Thus, although an appropriate physical environment and a strong mental inheritance may be essential to the rise of a great civilization, these two conditions may be at their best, and yet savagery may prevail because certain cultural essentials such as iron or beasts of burden are lacking.

From this example of people who lived in one of the finest climates but failed to make progress, let us turn to a case of quite the opposite type. The white settlers in northern Australia live in an almost tropical climate, which is poor as climates are rated in this book. Nevertheless, they maintain a high civilization and believe that they can raise that civilization still higher. They are sometimes cited as proof that climate in itself has little or no effect on health and that high civilizations may arise regardless of climate. Nowhere else in all the world does so large a body of white people, chiefly of British descent, live within the tropics. Some two hundred thousand upon whom the sun shines vertically at certain seasons are doing all the work of life with their own hands. The men work in the fields, on the docks, and among the cattle and sheep; they cut sugar cane under the tropical sun; but their health does not appear to suffer. The women work in intolerably hot kitchens, often among swarms of pestiferous flies, and in rooms where the blazing sun on the unceiled roof joins with the kitchen fire to produce an appalling temperature. Yet their death rate is phenomenally low, and they give birth to unusually healthy children. In fact the death rate in tropical Australia among all classes of the population is one of the lowest in the world.

In order to make fair comparisons between regions where the proportion of young people is unusually large and those where the proportion is smaller, let us use what is known as a "standard population." On this basis, which is universally accepted as the fairest yet devised, the death rate in Queensland, the most northerly state of Australia was 12.1 during the normal years before the World War. Compare this with 10.5 in New Zealand which is the healthiest country in the world, 12.4 in Denmark the healthiest country in Europe, 15.2 in England, and 16.1 in the part of the United States then included in the registration area.

In addition to this the medical examinations of young men who enlisted during the World War showed that the Queenslanders, even the tropical Queenslanders, were certainly not less robust and vigorous than those from other parts of Australia and much more so than were the similar men from England. Moreover, children of the second and third generation have been born in tropical Australia and are fine, active specimens. It seems to be proved beyond question that the white man can live in tropical Australia, that he can enjoy good health, and that white children of high vitality can be born there, live there all their lives, and become the parents of other children who seem to be equally healthy.

These facts are highly important and encouraging. They hold out a definite and not unreasonable hope that some day a branch of the white race may live permanently within the tropics and carry on all the work of life without the help of any colored race. The fond hope of the Australians that there may permanently be a "White Australia" is by no means without foundation. Does this invalidate our hypothesis of the relation between climate and civilization? Let us see what the facts really are. The Australians are so much interested in the problem of a White Australia that their Commonwealth Statistician has issued a special bulletin on tropical Australia. He has also kindly placed at my disposal certain other unpublished data. I am afraid that I shall seem to the Australians to use these data in a way that they will not like. But after all, what they really want is the truth. I am perhaps as keen as they for a White Australia; not only do I want to see my own race show its ability to conquer every kind of environment, but I am extremely anxious to see a conclusive tropical experiment on a large scale. I want to know beyond question whether white men can live permanently in a tropical country, conquer the diseases, and make progress without the help or hindrance of any colored race. Tropical Australia affords by far the greatest opportunity for such an experiment. It is peculiarly free from tropical diseases, and there seems to be no good reason why it should fail to be kept free; its colored inhabitants are few in number and most of them belong to the aboriginal Australian race which seems to be dying out and which is quite easily kept separate from the whites; and it is under a government which is disposed to do everything possible to make the experiment a success. I would give a great deal to be able to see what kind of people live in tropical Australia one or two hundred years hence. But if the tropical Australians of the next few generations are to succeed—indeed, if they are to avoid great misery perhaps—they must know exactly what they are facing. To let the dream of a White Australia and the hope of developing a great country blind ourselves to the facts is sure to produce evil.

We have already seen how extraordinarily low is the death rate of Queensland. If we take only the part of Australia that lies within the tropics we find that during the years 1920, 1921, and 1922 the following remarkable conditions prevailed.

	Australia as a whole	Non-tropical Queensland	Tropical Queensland
Birth rate per 1000 women aged 15-44	107	114	141
Infant death rate per 1000 births	63	56	55
Female death rate per 1000	8.66	8.03	7.90
Male death rate per 1000 persons	11.03	10.61	14.00

In the first three of these conditions non-tropical Queensland is more healthful and vigorous than Australia as a whole, while tropical Queensland is still more healthful and vigorous. In other words the white women in the tropical part of Queensland enjoy remarkably good health; they give birth to a relatively large number of children; and those children are uncommonly healthy. What more could one ask as a recommendation of the climate of a country? The men, to be sure, have a much higher death rate than those of other parts of Australia, but from the standpoint of the future of the race they are much less important than the women and children. The *Census Bulletin on Tropical Australia* attempts to explain the differences between the death rates of men and women as follows: "As both sexes should be affected by tropical conditions to something approaching an equal degree, it is apparent that there must be some cause, apart from purely tropical attributes, to account for the unfavorable male death rate. . . . The great excess of men over women in the tropical population suggests that there are many men living under primitive conditions. Under such circumstances there is too frequently a disregard for precautionary sanitary measures, and a reckless neglect of the first symptoms of disease. There are other cases, too, where it is impossible by reason of great distance to procure medical or surgical assistance. It would appear to be mainly to such circumstances as these that the high male death rate in the tropical parts of Queensland is due. Otherwise, the great difference between the mortality of the sexes is not readily explained."

There may be some truth in this, for in tropical Queensland there were in 1921 about 133 men for every 100 women. Nevertheless it must be a minor factor. Not only is it generally believed that frontier life is far more healthful than city life, but much more than half the men of tropical Australia, as well as a still larger proportion of the women, live in the towns where there are physicians and often hospitals. Moreover, as appears in the last sentence of the quotation from the census bulletin, the isolation of the men on the frontier is appealed to as an explanation of their high death rate only in the absence of any other reasonable cause.

There is, however, another reasonable cause of the low death rate among the women, a cause which the Census recognizes as applying to the population as a whole, but does not recognize as applying to women more than to men. That cause is natural selection. As the bulletin on tropical Australia puts it: "Perhaps the most potent influence [in causing the low death rate of tropical Australia] has been that sort of natural selection which operates in the settlement of all new territories. It may be stated, as a rule, that only the bolder and more virile of any community will venture on the rôle of pioneer in new unsettled country, and, when such a new country is a tropical one, to which popular opinion generally ascribes more than the usual discomfort, the physical standard of the settlers is probably more than ordinarily high."

The logical corollary of this statement is that among women the selection of those who are physically fit is much stronger than among men. Young men rarely think of their health; young women think of theirs a great deal, and between men and women 30 to 50 years of age the contrast is still greater. That the women distribute themselves over tropical Australia in a different way from the men is obvious from the fact that in the tropical part of western Australia, which is the most inaccessible and undeveloped, there were in 1921 about 524 men for every hundred women; in the tropical part of the Northern Territory, which is not quite so remote, the number of men per hundred women was 271; in tropical Queensland, which is still better, 133; and in Townsville, one of the two largest cities, 108. On the other hand, in Rockhampton, which is the largest city of tropical Australia and also the one located farthest south and hence in the least enervating climate so far as the coast is concerned, the number of men per 100 women was only 94. In other words the women become relatively more and more numerous as one gets away from two conditions, namely the frontier and a truly tropical climate.

But this does not end the matter. The women tend not only to concentrate in the best parts of tropical Queensland, but to remain out of tropical Australia altogether, or else to go away after once getting there. This is evident from the fact that

among children up to the age of about fifteen years the relative
numbers of boys and girls are normal. From that age onward
the number of men in proportion to women steadily increases.
For example, among young people 15 to 19 years of age there
are 110 boys for 100 girls, but at the age of 30 to 34 years this
proportion has increased to 155, while between the ages of 65 to
69 it rises to 249. Yet in a normal population at this latter age
there are only about 89 men for 100 women. Where such conditions prevail it needs no demonstration to prove that the
women either have never come to tropical Australia or have
come and gone away. In either case it is certain that a large
number of the women are missing because of their health. One
has only to talk with those who are still there, or with the
Australian physicians, to learn how often the men like the climate while the women dislike it, and how often the health of a
wife or daughter causes a family either to move away, or to be
temporarily separated. The women frequently go to the cooler
South, while the men stay in the tropical North. Often the whole
family moves away from tropical Australia because the women
do not like the climate. The selective process which causes the
men in a new tropical country to be especially strong and vigorous works still more effectively in causing the women and hence
the children to be even more markedly of the same type. The low
death rate in tropical Australia does not indicate that the
climate is favorable, but merely that natural selection has been
unusually vigorous.

Now look at the matter in still another way. Here is a little
table showing the death rate among persons from 15 to 49 years
of age in the three Australian states of Victoria, which is the
most southerly and hence the coolest, New South Wales which
also lies quite far to the south, and Queensland which lies so far
to the north that a quarter of its population is within the
tropics. The death rate has been calculated according to a
"standard population" so as to eliminate any differences arising

from the fact that there are more young people in one section than in another. The residents of the three states have been divided according to their birthplace, the great majority of Australians being natives of one of the three states, or else of England or Scotland.

DEATH RATES PER "STANDARD POPULATION" AT AGE 15-49 YEARS IN VICTORIA, NEW SOUTH WALES, AND QUEENSLAND ACCORDING TO PLACES OF BIRTH AND RESIDENCE, 1920, 1921, 1922.

	Residence								
Birthplace	*Victoria*		*New South Wales*		*Queensland*		*Aver.*		*Grand Av. both sexes*
	Male	Female	Male	Female	Male	Female	M.	F.	
Victoria	4.38	4.01	3.60	3.38	4.28	3.72	4.09	3.70	3.90
England	3.65	3.87	3.78	3.39	4.66	3.99	4.03	3.75	3.89
Scotland	4.83	4.13	3.94	3.48	4.85	3.47	4.54	3.69	4.08
New S. Wales	4.47	4.27	3.76	3.57	4.84	4.03	4.36	3.96	4.16
Queensland	5.22*	5.01*	4.42*	4.13*	4.87*	4.27*	4.84*	4.47*	4.66*
Average	4.51	4.26	3.90	3.59	4.70	4.00	——	——	——

Grand average for
the two sexes   4.39   3.75   4.35

The significant fact about the table is this: no matter whether they reside in Victoria, New South Wales, or Queensland, the people who were born in Queensland have a higher death rate than those born in the other two Australian states or in England or Scotland. If we combine the figures in the last column it appears that the residents of Australia who were born in Victoria, England, Scotland, and New South Wales have an average death rate of 4.01. On the other hand the death rate among the people born in Queensland, no matter where they reside, is 4.66 or approximately 15 per cent greater. The high death rate of Queenslanders in Victoria, or New South Wales,

* Maximum.

may be due to the fact that Queenslanders whose health is impaired migrate to those cooler states. But in that case the born Queenslanders who remain in Queensland ought to be stronger than the average and their death rate should be correspondingly low. On the contrary, their death rate exceeds that of the people born in any of the other four regions. In other words, although the people whose *residence* is in Queensland have a very low death rate, those who were *born* there have a high death rate. Those who reside there were largely born elsewhere, and have had the advantage of a strenuous though unconscious process of natural selection. Those who were born there are more nearly like the average of their race than were their selected parents, for that is the universal biological rule, and they show what seems to be the effect of the tropical climate.

Another significant fact in this connection is the number of children per family. We have already seen that in proportion to the number of women between 15 and 44 years of age the birth rate in Queensland is high compared with the rest of Australia, and is still higher in the tropical sections. That apparently is due to the fact that the weaker women either leave their husbands in the North and go to the cooler South, or else persuade their husbands to move away. Those who remain are strong and vigorous and have many children. But when we examine the number of children born to persons who are Queenslanders by birth, regardless of what part of Australia they later reside in, we find quite a different condition. The following figures show the number of children born to persons whose own birthplace was in various regions and who died in Australia during the year 1921.

Germany	6.3	Tasmania	5.0
Ireland	5.8	South Australia	4.5
Scotland	5.7	Victoria	4.1
England	5.4	Queensland	3.8
New South Wales	5.1		

The high birth rate among the foreign-born Australians is normal. There is nothing surprising about the differences among the birth rates of people born in New South Wales, Tasmania, and South Australia. The low rate among natives of Victoria, however, is not easily explained unless it be that Victoria, more than any of the other states, has lost the pioneer, frontier quality, and hence has the relatively low birth rate which is normal in old regions. For Queensland, no such explanation can be advanced, for that state has more of the pioneer quality than any other in our table. The most probable explanation is that the birth rate is low for the same reason that the death rate is high. Both conditions indicate that in spite of the strong selective process to which their parents were subjected, the native-born whites of Queensland display qualities which suggest a lower degree of physical stamina than prevails among the other Australians, and which seem to be the result of the tropical climate. If this is the case it strongly bears out the general thesis of this book.

Even the part of Australia where the tropical Queenslanders live is not tropical in the ordinary sense. The climate is indeed quite trying by reason of the length and monotony of the summer and because of the dampness near the coast and the great heat in the interior. There is, however, a real winter in the regions where most of the tropical Queenslanders actually live. For two or three months the temperature almost everywhere falls to between 40° and 50° during the night, while on the highlands there is often frost. If a climate so mildly tropical can produce such clear results, it seems probable that a thoroughly tropical climate may handicap the white man to a far greater extent. Nevertheless, I do not despair of the ultimate triumph of the white man over the tropics, and much less do I despair of a White Australia. The process of selection for climate has never been really tested. All that has happened thus far has been purely fortuitous. The hope of the future, I believe, lies in an orderly and far-sighted selection of the right types of people, as well as in the further development of tropical medicine and hygiene.

But suppose the matter of health should be taken care of to such a degree that the white race could live, thrive, and permanently reproduce itself within the tropics, would civilization there advance as in cooler regions? I doubt it. In the first place the most casual observation in any tropical region shows that the white man as well as the colored does not act with so much energy as in cooler climates. No matter how good his health, he is forced to work comparatively slowly or for short hours, or else he exhausts himself. It was a Queensland sheep "selector," the owner of twenty or thirty thousand animals, who first called my attention to the "Queensland walk." We were watching a man who slowly sauntered from the hotel to the railroad station. "That's the way we all walk after we've been here in tropical Australia a while." Not only do people almost of necessity walk that way, but they often work that way. In tropical Australia, just as in other tropical countries, all sorts of things are allowed to lie at loose ends much more than among the same kind of people in more stimulating climates.

The new point which I wish to make in regard to Australia is that natural selection works against a tropical country, and is working that way even more now than in the past. This sounds like a contradiction to what has just been said about the strong physique of the Queenslanders and the possibility of building up a genuine tropical race of white men. But there is no contradiction. I am talking now about the qualities which lead to progress in civilization. As one goes about in tropical countries and especially in Australia, one soon finds that the people who complain most about the climate are, first, the women, and then the professional men. In other words it is the people whose work keeps them indoors. The rancher, planter, promoter, real estate agent, cane cutter, and dock laborer do not seem to feel the heat half as much as do the minister, lawyer, doctor, engineer, and teacher. These intellectual workers may and do adapt themselves to the climate and achieve results of high importance. Nevertheless, almost without exception, they are on the watch for a chance to get away to a pleasanter and more stimulating climate. Many of them are ready to accept smaller salaries for the sake of being in an atmosphere which is physically and mentally more stimulating. In this fact, perhaps, lies one of the greatest hindrances to the development of a high civilization in countries with unfavorable climates. We have already seen this selective process at work in the Bahamas, it is very active in tropical Australia, and it is energetically at work in the southern United States. The greater the wealth of a country, the greater its intellectual and physical activity, the better its transportation system and the more fully it picks out and educates all its bright sons and daughters, the more likely those same sons and daughters are to move away from the unstimulating environments and into those which most fully give scope to all their faculties. In this respect, as in so many others, each new advance of knowledge, each new step in the mastery of nature, gives greater power to the regions whose climatic advantages already help them to be leaders. To him that hath shall be given, and from him that hath not shall be taken away even that which he seemeth to have.

CHAPTER XVIII

THE CLIMATIC HYPOTHESIS OF CIVILIZATION

WE come now to the final step in our study of the hypothesis that climate ranks with racial inheritance and cultural development as one of the three great factors in determining the conditions of civilization. As to which of the three is most important it is impossible to say. The absence of good conditions in any one respect may hold a country back, while all three must rise to a high level if a race is to reach the highest plane of civilization. In this last chapter I propose to consider one final objection to the general hypothesis that climate has been a determining factor in the geographical distribution of human progress. That objection is that it seems impossible that changes of climate can have exerted so great an influence as is claimed in this book.

The first and most obvious effects of climatic changes are economic. We have said little about them in this book because they have been fully discussed in the publications numbered 2, 3, 7, 8, 10, 12, 14, 15, 17, and 22 in the list at the end of the preface. A single concrete case will illustrate the matter. In 1909 I visited Palestine. That happened to be an unusually dry year. During the months of April and May I rode scores of miles on horseback amid fields of wheat that showed merely a few scanty stalks three to six inches high. In Beersheba I talked with a number of men who had tried to raise grain during a few preceding years of good rainfall, but were financially ruined by the drought of 1909. In Moab many villages which had been recently reoccupied under the stimulus of the building of the

Mecca railroad were being abandoned. The inhabitants were moving back across the Jordan to the better watered parts of the country. Coincident with this migration of the agricultural people away from the drier areas along the borders of the desert, there was a corresponding outward movement of the Arabs from the desert. Because of the drought the great tribes from the interior of the Arabian desert swarmed over the grain fields on the eastern and southern border of Palestine, and let their camels eat up the few scattered stalks of wheat that had survived. The poor villagers shot at the Arabs when they first appeared; then ran in terror to the villages. Soldiers were sent to stop the Arabs, but in vain. The quarrels between the Arabs and the peasants resulted in so many gunshot wounds that Dr. Patterson of the Presbyterian Hospital at Hebron was swamped with patients. I myself was in an Arab encampment when it was raided by other Arabs. I was held up and my money demanded at another time, and I came in personal contact with raiding parties on three other occasions. Thus a single year of drought brought distress upon all kinds of people; it caused migrations among those of the settled population who occupied the least favored locations; it set the desert tribes in motion; and it spread poverty, hunger, raids, war, wounds, and death on every side.

Consider now what must have happened when Palestine changed from a condition of relatively abundant rainfall in the early part of the Christian era to a condition drier than that of today in the seventh century. Remember that at certain periods, such as the third century B.C. and the beginning of the seventh century A.D., a relatively rapid change took place. Remember too that Professor Butler of Princeton has been quoted above as saying that an area of "20,000 square miles of land on the eastern border of Syria is now too dry for permanent human occupation, but was once more thickly populated than any area of similar dimensions in England or in the United States outside the immediate vicinity of the large modern cities." This must mean a population of at least 100 per square mile or 2,000,000 people. South of the region described by Butler there are additional thousands of square miles of similar dry land full of ruins, while further east, in northern Mesopotamia, in the region described above by Clay, the same is true. Suppose for the sake of argument that 3,000,000 people lived in these areas which are now too dry to be occupied.

When a diminution of rainfall gradually rendered this great multitude homeless, a corresponding reduction must have taken place in the productivity of the areas which are still inhabited. Palestine today contains about 650,000 people and Syria about 3,000,000. There is abundant evidence that when the outlying districts were habitable the population of the still habitable areas was much greater than now. It seems, then, that at the time of Christ, or at least some 400 years earlier, there may have been over 6,000,000 people in these areas of Palestine and Syria. Then what happened? Regions which had been supporting perhaps 9,000,000 people in comparative comfort were so stricken by drought that the homes of about three million became almost uninhabitable, while the productivity of the remainder was reduced nearly half. Thus only about three and a half million people could live where 9,000,000 had lived before. Of course so great a decline did not occur in a single generation or even a single century. But a reduction of the capacity of the country from 9,000,000 to 8,000,000 persons would produce misery, famine, death, migration, raids, wars, and misgovernment to an almost incredible degree. Then would follow a few decades or scores of years of recovery and again a reduction in the capacity of the country to support people. Each such blow must have dragged civilization downward, and the resultant chaos of the Dark Ages is by no means surprising.

The misery and discontent due to prolonged poor crops tend to make people unstable not only politically, but in other ways.

Religious bitterness is almost sure to increase under such conditions. A portion of the community attributes its poverty to the fact that there is something wrong with the present form of religion. The rest are inclined to attribute their distress to the wickedness of their neighbors who decry the old religion. Thus bitterness and persecution are engendered. Those who become discontented with the old religion are prone to accept the new ideas propounded by religious enthusiasts. This seems to have been the case when Mohammed made his appeal to the Arabs after the prolonged period of increasing aridity which culminated with a sudden access of dryness in the first half of the seventh century. Without the genius of Mohammed that long period of adversity might have come to an end without any serious upsetting of the old conditions; but without the discontent and unrest fostered by years of distress Mohammed might have appealed in vain, for he would have had to speak to men who did not desire change, instead of to those who ardently longed for it.

The people of the desert may perhaps occasionally pour forth from their homes without any special stimulation, but this is doubtful. Anyone who has had much to do with the Arabs and other desert nomads knows that when there is plenty of water and grass there is very little raiding and fighting. On the other hand, a single dry year causes raids in the fashion described above; and prolonged dry periods appear to lead to great outbursts of desert people like that which reached its crest under the influence of Mohammedanism, although it began before Mohammed came into prominence. The Mohammedan migration was by no means unique. Many of the barbarian migrations of earlier times seem to have been impelled by similar dry periods.

At a later time two similar invasions took place. First, from about 1000 A.D. to 1200 A.D. the climate of central Asia seems to have grown decidedly drier, and distress and discontent reigned among the people of the tents. At this time, as in the days of Mohammed, no great concerted movement might have arisen, had it not been for the ambitions of one man. Jenghiz Khan may have been no more ambitious and no abler than other gifted men of his race, but he happened to live at a period when natural calamities had brought his people to a condition of discontent favorable to his aspirations. This condition, it would seem, was an important help in enabling him in a few years to arouse all the tribes of the steppes and deserts, and sweep over Asia, bringing almost unparalleled devastation. Three centuries later, at the beginning of the sixteenth century, another ambitious Asiatic, Baber by name, arose in Turkestan and emulated his great ancestor Jenghiz Khan. In Baber's case, also, physical conditions seem to have favored his projects, for after a period of improved climate, a rather rapid decrease in rainfall culminated at the time of his conquests. How much this had to do with the fact that he was able to conquer India and establish there the Mogul dynasty has not yet been thoroughly investigated. It seems clear, however, that this dry period should at least be carefully considered before any conclusions are drawn as to Baber and the Mogul conquest of India.

From great wars and movements of the nations let us turn back to the individual people, and see how increasing aridity may effect a race physiologically. The chief effect is probably produced by the selective action of disease. Insidious diseases such as malaria, consumption, neurasthenia, and the like are presumably among the most important sifters of the wheat from the chaff in the physical make-up of a nation. Although malaria, for example, does not kill people in any such spectacular fashion as do the great epidemics, it may be far more dangerous in its ultimate effects. The plague passes over the land and is gone; the dead are dead, and the living have suffered no serious injury. Malaria on the contrary, hangs on year after year, not killing its victims, but sapping their energy and vitality. The presence and the abundance of malaria are closely associated with climate and topography. Without entering into any discussion of the origin of malaria, let me point out how a change toward aridity in a country like Greece and, to a less extent, Italy, would probably foster the disease.

Malaria is preëminently a disease of tropical and subtropical countries whose climate is characterized by alternate wet and dry seasons. Except in the perennially moist portions of the tropics, the streams of such regions are subject to seasonal floods which spread over wide areas for a short period and then disappear, leaving innumerable stagnant pools and swamps, ideal breeding places for the anopheles mosquito. Permanent bodies of water usually contain fish which eat the mosquito larvæ and reduce their numbers, or else the water moves sufficiently to carry away most of the eggs that are laid in it. When the climate of a subtropical country becomes drier, the conditions which favor the mosquito are intensified. This is due in large measure to the fact that a diminution of the rainfall lessens the amount of vegetation upon the slopes and thus allows the soil to be washed away rapidly. The streams are thereby overloaded and begin to fill the valleys with sand and gravel. This causes the flowing water to wander hither and thither over broad flood plains in innumerable channels, which form pools when the floods assuage. Or it may be that the water loses itself in marginal swamps. The streams also become intermittent, and no longer contain large quantities of fish. Thus everything coöperates to reduce the number of streams which flow steadily throughout the year, and to increase the number of bodies of stagnant water in which the mosquitoes may live. This in itself may produce most widespread effects. How great they are may be judged from the success of the United States government in eradicating malaria at Panama by the opposite process of reducing the number of places where mosquitoes can breed.

At the present time malaria is endemic in Greece and Rome. That is, it is always there, and is looked upon as one of the necessary diseases of childhood, much as we look upon measles. Sir Ronald Ross of the Liverpool School of Tropical Medicine is responsible for the statement that nearly half the people of Greece have suffered genuine injury from malaria, and in Italy the case is scarcely better. Up to the age of puberty children are attacked by it every autumn. They grow weak and sallow, their spleens are permanently enlarged, and their vitality is lowered for life. No one who has known much of malaria will question the severity of its results and the length of time which elapses before they are eradicated even in the case of adults. In spite of quinine, which has come to our aid in modern days, malaria is one of the most insidious of diseases. Every traveler who is really familiar with the Orient knows how the sufferers from malaria lie and groan for days, and may have little energy for months. They go languidly to the necessary tasks, and as soon as possible sit down to rest with open, stupid mouths. Physicians agree that it is impossible to expect much initiative or energy from a nation in which for centuries almost half of every generation has been devitalized by this baneful disease.

From a painstaking study of classical authors W. H. S. Jones* has concluded that up to about 400 B.C. in Greece and 200 B.C. in Rome, malaria was almost unknown. Then it appeared, and during the succeeding century or two became common. At first it attacked adults, which shows that it was a relatively new disease, which was still epidemic and not endemic, or else, we would add, that Greece was on the very border of its habitat. Later it became permanently located in the respective countries and attacked chiefly children, the older people having become immune after suffering in childhood. It is noticeable that the introduction of malaria coincides with the beginning of the weakening of Greece and Rome, and the time when it became

* Malaria: A Neglected Factor in the History of Greece and Rome. Cambridge, England, 1907.

endemic, in Greece at least, is synchronous with the epoch when the luster of the ancient names became irretrievably dimmed.

Ross and Jones are of the opinion that, along with various other factors, malaria was one of the important causes of the fall of Greece and Rome. The growing effeminacy and lightness of the Greeks, and the brutality of the Romans, are just the effects which they think would be produced upon people of the respective temperaments of the two races. The case is so strong that one can scarcely resist the conclusion that this pathological factor may have played an important part in the psychological changes which appear to have accompanied the decline of civilization and of population in both Greece and Rome. It would be unwarranted to assert that the increase in the amount and severity of malaria was due wholly to climatic changes. Other influences, such as contact with Egypt and the introduction of slaves, may have been equally effective. Nevertheless, it seems probable that the spread of the disease in both Greece and Rome proceeded most rapidly when a change of climate not only rendered the topography of the valleys and the behavior of the streams more favorable than hitherto to the propagation of the anopheles mosquito, but likewise weakened the physique of the people so that they readily succumbed to disease.

Natural selection presents a still more insidious way in which the change from relatively moist, stormy, cool conditions to those of aridity may have affected the Greek, Roman, and other races. In the opinion of many scholars one of the most important factors in the greatness of these powers was the presence of a race of blond northern invaders. Take the case of Greece. These northern Achæans apparently came into the country in the thirteenth century B.C. Their coming may have been influenced by the dry period of which we find some evidence at that time both in America and Asia. After their arrival the climate on the whole, although with many fluctuations, appears to have become more propitious. Up to the third century it continued to be favorable. Then it became more arid. We have seen how sensitive people are to climatic environment. The negro would apparently disappear in the northern United States were he not replenished from the South. The Scandinavian does not seem to prosper greatly in the dry, sunny portions of the United States; he is there subject to diseases of the skin and nerves which appear seriously to deplete his numbers in a few generations; whereas in the rainy Northwest, which resembles his native habitat, he thrives greatly both in body and estate. It was probably the same with the northern invaders in Greece. So long as the climate was propitious they flourished and lent strength to the country. Then, when conditions became less favorable, the unseen ravages of malaria and other diseases presumably attacked them with especial severity, and in the course of centuries they gradually disappeared. Today blond Greeks are almost unknown, although classical literature and many fair-haired old statues demonstrate their presence formerly in considerable numbers.

Thus far we have been discussing changes of climate, and have overlooked the fact that even though the climate remains constant, man's cultural progress may alter the location of the most favorable climate and hence of the centers of civilization. This depends upon the fact that the climate which is most favorable to a race in a primitive stage of development is not the most favorable for the same race in a higher stage of culture. GilFillan, in an article on "The Coldward Course of Progress,"* has presented an interesting study of the way in which man's cultural heritage, in the form of a constantly growing command over nature, has enabled mankind to advance farther and farther into regions of low winter temperature. In a savage state, without fire, clothing, or shelter other than caves, man is not likely to thrive where the winter temperature remains for any length

* *Political Science Quarterly*, 1920. The same subject is discussed from another standpoint in Stefansson's *The Northward Course of Empire*.

of time as low as freezing, and his main development is almost sure to be where the coldest months are not much below the physical optimum of 64° or 65° F. With each step of progress he is able to endure greater extremes of climate. The open fire, the grate, the stove, the furnace, and the central heating plant are all steps toward a condition where man is able to resist the cold, no matter how severe it may be. The use of the skins of animals for clothing; the invention of the arts of spinning and weaving; the domestication of the sheep, goat, and camel; the cultivation of flax, and especially cotton; the invention of power looms and of the steam engine;—all these are important steps which have tended to make man independent of climate. So too are the inventions which enabled mankind to close the mouths of the caves in which he lived, and then to build huts, thatch them, cover them with close-fitting shingles, construct them of thick bricks, or otherwise make them warm and weatherproof. In like manner man's power to resist unfavorable climates has been increased by every invention that has made it possible to transport food cheaply and quickly, and to preserve it for long periods. Today it is not difficult for large communities to live in health and comfort in the cold parts of the earth where their naked ancestors would have perished in a few days. Human energy may perhaps ultimately rise higher in those regions than in others, for many of the diseases of warmer regions do not flourish there, and the constant changes from the warmth within-doors to the cold without seem to be highly stimulating and healthful, provided they are properly guarded against. Moreover, in cold regions man can be free from the effects of undue heat against which he has thus far shown little ability to protect himself. Thus it appears that even without changes of climate the highest civilization would have tended gradually to migrate farther and farther north. How far this tendency will go is not yet clear, but the far northern location of such progressive people as the Scotch, Canadians, Swedes, Norwegians,

Icelanders, and Finns seems to suggest that the limit has not yet been reached.

One other phase of GilFillan's work deserves careful notice. He points out that when civilization takes a backward step it also tends to move equatorward. For instance, about two hundred years before Christ the leadership of the world had passed from Greece, while Rome was falling into a state of confusion and retrogression from which its recovery was never complete, even in the days of Augustus. At that time the intellectual leadership went back toward the south, and the Alexandrines held aloft the torch of learning. They did not, to be sure, make any great intellectual or material innovations, but they at least kept the spirit of progress alive. In the same way in the Dark Ages the intellectual leadership of the world went back to North Africa where lived such men as Saint Augustine and his confreres.

GilFillan suggests an interesting explanation of what happens when the most progressive nations of any given period are suddenly thrown backward by barbarian invasions, climatic changes, or similar occurrences. At such times regions which have formerly ranked high stand out once more as leaders. When the cooler countries fall to a low estate, the eclipsed warmer countries come into prominence. They still represent the highest type of civilization that is compatible with their climate, even though they have not been able to make that type appear great or important because the cooler and more energetic countries have not given them a chance. Thus in the Alexandrine period, when Greece had fallen and even Rome was at a low ebb, northern Egypt took the lead because the stage of progress to which the world had regressed was suited to Egypt's climate. How true this hypothesis may be, it is hard to say. The migration of the more competent Greeks to Egypt may explain the Alexandrine period. Greece was becoming uninhabitable and poverty stricken by reason of an adverse climatic change, but

Egypt with its perennial river was still prosperous and attractive. Nevertheless, there seems to be much truth in the idea that man's social progress constantly alters his relation to climate. In the past great inventions have helped chiefly in enabling man to overcome low temperature; in the future, perhaps, they will help him in equal measure to overcome high temperature, dryness, and monotony.

The last matter to engage our attention is the effect of changes in storminess. In reading what has been said about Egypt, for example, the reader has probably said to himself: "This book claims that the people there were once energetic and healthy because of frequent storms. If this is so, the rainfall must have been much more abundant than now. For the sake of argument we grant that at certain periods the precipitation was greater than at present, but we cannot possibly believe that it increased to the extent demanded by this hypothesis."

This objection is so important that I have made a special investigation to determine the exact relation between storminess and rainfall. The word storm, as already explained, does not primarily mean rain. To the meteorologists a storm is an area of low barometric pressure which is always accompanied by inblowing winds, and usually, but not invariably, by rain. Even if there is not a drop of rain, a storm may otherwise be fully developed and may cause strong winds which give rise to changes of temperature and humidity. The matter is well illustrated by comparing Colorado and Georgia. According to Kullmer's maps, Colorado is one of the stormy parts of the United States. It is crossed by the centers of about three times as many storms as Georgia. Yet its rainfall is only about one third as great. In other words, Georgia has nine times as much rain in proportion to its storms. The reason is simple. Colorado lies far from the ocean, and the air which rises in the centers of its areas of low pressure has already lost most of its moisture. Georgia, on the other hand, lies so near the broad Atlantic and the warm Gulf of Mexico that storms draw in great quantities of moisture. The lands around the eastern Mediterranean and in western Asia have a climate far more like that of Colorado than of Georgia. Part, to be sure, are near the Mediterranean Sea, but that furnishes only a little water compared with the great oceans.

Another comparison shows that in the entire United States if the number of storms increases 20 per cent during a given series of years, the rainfall increases scarcely 10 per cent. In the arid southwest, however, the ratio is larger, the increase in storminess for all available stations being three times as great as the increase in rainfall. These figures apply to present conditions where the variations are slight and of short duration. If larger, more permanent changes took place, the rapidity of the atmospheric circulation would probably be so much increased that the ratio between increase in storminess and in rainfall would be greater than at present, and might be four, or even six to one. The drier the region, the greater would probably be the ratio.

Suppose that the ratio were four to one in Syria. A change which would increase the storminess by 200 per cent would involve a change of rainfall amounting to 50 per cent. That is, the number of stimulating changes would be three times as great as now, while the rainfall would only increase by one half. Such a change would render the Phœnician coast much more stimulating than at present. It would also increase the agricultural wealth, and would cause the limits of permanent habitation to advance some miles into the desert. There the stimulating influence might be less than on the coast, because the storms might be somewhat interrupted by crossing the mountains. Nevertheless, it would be important. The effect of increased storminess upon habitability, however, would be much less noticeable than upon energy or than in better watered regions. An increase of 50 per cent in a rainfall of twenty inches, such as is now enjoyed by many parts of Syria, would raise the precipitation to thirty inches, a figure which permits great prosperity. A corresponding change in Egypt would increase the rainfall of Alexandria from 8.8 inches to 13.2, that of Port Said from 3.3 to 5, and that of Cairo from 1.3 to 2. Farther east in Mesopotamia, Bagdad would change from 9 to 13.5, while in Persia, Teheran would rise from 10 to 15 and Ispahan from 5.2 to 7.8. Deserts would still be deserts. They would be easier to cross than at present, and the number of inhabitants might be greater, for there would be more pasturage for camels and sheep. The springs would also be larger and more permanent than now, and some new ones would appear. Yet the predominant feature would still be great wastes of blowing sand and barren gravel; the people would have to be nomads; and those who entered the desert would frequently encounter rigors like those of today. Thus a large change in the stimulating qualities of a subarid or desert climate is possible without a change of rainfall greater than that for which there seems to be good evidence.

Consider more specifically exactly what happens when storminess increases in regions such as western Asia and eastern California. The general conditions at the time when Greece was in its prime, for example, appear to have been approximately as follows: (1) The *average temperature* for the year as a whole was probably a little lower than at present, but it is doubtful whether the difference amounted to more than two or three degrees at most. (2) The temperature of the *seasons* may have varied somewhat more than that of the year as a whole. The greater mixing of the air from wide areas by means of the winds that accompany storms presumably lowered the summer temperature a little over the lands and raised the winter temperature. (3) The amount of *rain* was apparently considerably greater than now. Doubtless the rain then, as now, came chiefly in winter, while the summers were very dry just as at present, but the length of the rainy season appears to have been increased. That is, the rains began earlier in the autumn and lasted longer in the spring than at present. (4) The amount of *atmospheric moisture* over the lands was presumably greater than now. This would arise partly from the increased cloudiness and rainfall and partly from the fact that the winds were presumably more efficacious in bringing moisture from the neighboring seas. (5) The *winds* were apparently stronger than at present, and varied in direction much more than now because of the frequent storms. (6) The greater storminess and the more frequent changes in the winds, as well as their greater strength, must have caused great or at least frequent *variability of temperature* as well as variations in other respects. Such variability would occur even in summer when the storms passed to the north of the lands that were most progressive two or three thousand years ago; it would be still more pronounced in the rainy season.

Let us apply these generalizations to a specific case. Suppose that from 500 to 400 B.C. the climate of Athens differed from that of the present in the following respects:

1. Temperature of July 77° F. instead of 81° F.; January 48° instead of 46°. Mean temperature of the year 62.0° instead of 63.1°.

2. Relative humidity at all seasons 10 per cent higher than now, January the moistest month, 84 per cent instead of 74 per cent; July the driest month, 58 per cent instead of 48 per cent. As a matter of fact the change was probably not the same at all seasons, but I wish to keep our example simple.

3. Annual rainfall 22 inches instead of 15, ranging perhaps from 3.3 inches in November to 1.0 in July instead of 2.9 in November and .03 in July.

4. Number of storms twice as great as now.

Such a change seems conservative. It is scarcely more than

the normal variation from one year to another. Nevertheless, when we calculate its effect upon health in the same way that we have calculated the data for our map of climatic energy based on the effect of the seasons in American cities (Figure 36), the result is astonishing. From a climatic level only equal to that of Augusta in Georgia and Vicksburg in Mississippi, Athens rises to a level practically the same as that of New York and Chicago, the best regions in North America aside from the coast near Newport; a slight farther increase in storminess or a farther lowering of the summer temperature two or three degrees more would make Athens rival Paris and Berlin; and if the storminess should be three times as much as at present, which would be a small matter compared with the differences between one year and another at places like San Francisco, the healthfulness and stimulating qualities of the climate of Athens would rival those of southeastern England, which seems to be wellnigh the most favored place in the whole world. When we recall that according to Figure 41 the estimated death rate in Greece is nearly twice that of England, the significance of such a change is apparent. It would presumably get rid of malaria to a large extent; it would stimulate the Greeks to a degree of persistent activity quite foreign to the country at present; it would raise the economic level and correspondingly improve the diet of the people; and it would give to the Greeks a spirit of enterprise, a physical vigor, and a mental activity which would probably soon enable them to take advantage of all sorts of modern discoveries which they now use half-heartedly and ineffectively, if at all. The slightness of the climatic change necessary to produce important results seems to give to the conclusions set forth in the present edition of this book a reliability much greater than was possible in the first edition, before the health of American and European cities had been studied in its relation to the weather.

As we come to the end of this volume I am well aware that to those who accept the climatic hypothesis, it may seem depressing. To the dweller in the less favored parts of the world it may appear to sound the knell to his hopes for great progress in the land that he loves. To his brother in the center of modern activity a most disquieting vision of possible retrogression is disclosed. If our reasoning is correct, man is far more limited than he has realized. He has boasted that he is the lord of creation. He has revelled in the thought that he alone among created beings can dwell in the uttermost bounds of the earth. One more of the bulwarks of this old belief is now assailed. Man can apparently live in any region where he can obtain food, but his physical and mental energy and his moral character reach their highest development only in a few strictly limited areas. The location of those areas appears to have varied greatly in the past ; it may vary greatly in the future. In a thousand years, for all that we can tell,—so the prophet of evil will say,—no highly favorable region may exist upon the globe, and the human race may be thrown back into the dull, lethargic state of our present tropical races. Even without so dire a calamity, the location of the regions of greatest climatic energy may in a few hundred years change again to Egypt, Mesopotamia, and Guatemala. The consequent rise of new powers and the decline of those now dominant may throw the world into a chaos far worse than that of the Dark Ages. Races of low mental caliber may be stimulated to most pernicious activity, while those of high capacity may not have energy to withstand their more barbarous neighbors.

Even if such extreme disasters should not occur, the prospect is depressing. Take such a favored country as the United States. In the South we find less energy, less vitality, less education, and fewer men who rise to eminence than in the North, not because southerners are in any way innately inferior to northerners, but apparently because of the adverse climate. In the far West people seem to be stimulated to such a degree that nervous exhaustion threatens them. In the North we see still another handicap. In spite of a wonderfully stimulating climate most of the year, the people suffer sudden checks because of the extremes of temperature. These conditions favor nervousness, and worst of all they frequently stimulate harmful activities. That, perhaps, is why American children are so rude and boisterous, or why so staid a city as Boston has six times as many murders as London in proportion to the population. Our country takes immigrants of every mental caliber, and then stimulates some to noble deeds and others to commit murder, break down the respect for law, and give us city governments that shame us in the eyes of the world. All these things would apparently not happen to such an extent were our climate less bracing and did not its extremes often weaken the power of self-control. Other lands also have their drawbacks. Germany is much like the eastern United States, although not so extreme. France on the other hand is less stimulating. England suffers from too great cloudiness, and in Ireland this becomes a factor of serious import. If the best parts of the earth have such climatic disadvantages, what shall we say of Russia, weighted down with benumbing cold and comparative monotony or with changes so extreme that they are harmful? What of China under a much heavier handicap of monotony; or of tropical lands burdened most heavily of all? If climatic conditions influence character as we have inferred, does not our hypothesis weaken man's moral responsibility? Will not people more than ever ascribe their failings to nature, and so excuse themselves? In the favored regions will not men become increasingly arrogant and overbearing, because they will be surer than ever that the rest of the world cannot resist them? If all these sad results are possible, is it well to know that climate so strongly influences us? We cannot change the climate, so why ascribe to it such great effects merely to destroy hope in some and moral responsibility in others?

The answer to these questions may be put in the form of a parable. Ages ago a band of naked, houseless, fireless savages started from their warm home in the torrid zone, and pushed steadily northward from the beginning of spring to the end of summer. They never guessed that they had left the land of constant warmth until in September they began to feel an uncomfortable chill at night. Day by day it grew worse. Not knowing its cause they traveled this way or that to escape. Some went southward, but only a handful finally returned to their former home. There they resumed the old life, and their descendants are untutored savages to this day. Of those who wandered in other directions all perished except one small band. Finding that they could not escape the nipping air, the members of this band used the loftiest of human faculties, the power of conscious invention. Some tried to find shelter by digging in the ground, some gathered branches and leaves to make huts and warm beds, and some wrapped themselves in the skins of the beasts that they had slain. Soon these savages had taken some of the greatest steps toward civilization. The naked were clothed; the houseless sheltered; the improvident learned to dry meat and store it with nuts for the winter, and at last the art of making fire was discovered as a means of keeping warm. Thus they subsisted where at first they thought that they were doomed. And in the process of adjusting themselves to a hard environment they advanced by enormous strides, leaving the tropical part of mankind far in the rear.

Today mankind resembles these savages in certain respects. We know that we are limited by climate. As the savages faced the winter, so we are face to face with the fact that the human race has tried to conquer the arctic zone, the deserts, and the torrid zone, and has met with only the most limited success. Even in the temperate zone he has made a partial failure, for he is still hampered in hundreds of ways. Hitherto we have attributed our failure to economic conditions, to isolation and remoteness, to racial incapacity, or to specific diseases. Now we see that it is probably due in part to lack of energy or to other unfavorable effects produced directly upon the human system by climate. This is no reason for despair. We ought rather to rejoice because, perhaps, we may correct some of the evils which hitherto have baffled us.

Again and again in our discussion of factories and other matters we have come upon ways in which a change in our methods may do much to overcome the harmful effects of climate. I do not propose to enumerate them, for the specific application of our results may well be deferred until we know whether our main hypothesis is likely to stand. Yet one or two general lines of progress may properly be pointed out. Take the harmful winters of the northern United States. It is highly probable that the loss of energy which occurs at that time may be largely avoided, or at least greatly diminished. Much of it arises from the fact that after the wonderfully stimulating autumn weather, when we have been living under almost ideal conditions of mean temperature, of humidity, and of variability from day to day, we suddenly begin to heat our houses. We create an indoor climate of great uniformity, of unduly high mean temperature, and of the most extreme aridity. All these conditions are harmful. If our houses were kept at lower temperatures, if the temperature were varied from day to day, and if the humidity were kept at the optimum, we should increase our efficiency greatly. We should be comfortable, also, for with proper humidity, and with changes from day to day, we should not feel the need of the high temperatures which we now require because the extreme dryness forces the body to give up much more heat than would be demanded by air of greater humidity. Moreover, the uniform dryness within doors does almost untold harm in parching the mucous membranes and thus rendering us peculiarly liable to colds, grippe, and similar ailments which often lead to serious diseases such as pneumonia and tuberculosis. Of course we could not entirely avoid colds by the method here suggested, but we surely could diminish them. In the autumn before our houses are heated, colds are comparatively rare, and the same is true among people who live out of doors in winter. If the conditions inside our houses could be like those that prevail in the autumn, the general health of the community would probably be much improved. In this one way there might be a saving not only of millions of dollars' worth of valuable time, but of an immense amount of nervous energy which is wasted because persons who are irritated by colds do or say things that they would scorn under normal conditions.

Along still other lines great improvements might be possible. For instance, in many factories the same amount of work is expected each month. Hence, at certain seasons many operatives, especially girls, work harder than they ought, while at others they do not work so hard as they easily could without special effort. If factories were run in accordance with a well-established seasonal curve of energy, we should find the machinery running slowly in winter, faster in the spring, and in May perhaps 10 or 15 per cent faster than in January. Then in the summer it would run more slowly than in May, but not so slowly as in winter. Finally, in the autumn it would run at greater speed than at any other time of year. The operatives would scarcely be conscious of the difference, and they would probably do more work and preserve their health better than under the present system.

If our hypothesis is true, it is likely to prove helpful not only to places where the climatic disadvantages are slight, but where they are great. Consider regions which have a winter of great severity, but an invigorating summer. Contrast them with places where the summer is too hot, but the winter favorable. Russia and Mesopotamia may serve as examples. Today we already have a small number of people who move back and forth each year between places of this sort, for instance northern

Germany and the Riviera, New England and Florida. Unfortunately, those who do this are usually not the workers, but the idlers or those whose work is almost finished. In the future, however, if the principles here laid down find acceptance, we may expect that such interchanges will take place on a scale to stagger the imagination. Not only the leisure classes, but laborers and farmers may thus move back and forth. In winter most of Russia's peasants have little to do, and their enforced idleness is harmful. They might go to Mesopotamia where most of the farm work is done between October and May. If people could move thus from place to place, not only would there be an enormous increase in production, but many other benefits. The part of the population that moved would be stimulated, not only by the change of climate, but by contact with other races and new methods. They would be more tolerant, more progressive, and more eager for education. Both countries would benefit by such an interchange of workers, and much of the handicap of places like Russia might possibly be overcome. Perhaps the day will come when only the poorest families will stay in an unfavorable climate more than a few years at a time without going at least for a season to some place where they will receive new stimulus.

In tropical countries the chances for improvement are at a maximum. Already most Europeans and a few natives appreciate the necessity for spending part of the year among the mountains or in a climate different from that where they usually live. For the most part the lowlanders go to the highlands, but in lofty plateaus like Mexico it is not uncommon for foreigners to take a run down to the coast for a change of air. Of course altitude has much to do with this, but even though Vera Cruz has a bad climate for permanent residence, it is stimulating for a short while when one comes from a wholly different environment. In the future we can scarcely doubt that this method of overcoming the evil effects of a tropical climate will be resorted to on a vast scale, not only by foreigners, but by the more intelligent portion of the natives.

In the warmer parts of the earth there in another side to the question. Mankind needs not only seasonal changes, but variations from day to day, or week to week. Two methods of obtaining these suggest themselves. One is by cooling the interiors of houses. Today this is done on a small scale by shutting out the sun and sprinkling water to cause evaporation. There is no reason why the same result should not be produced on a large scale. We already know how to cool houses as well as to heat them. We do it in ice-plants. A thousand years ago men would have laughed at the idea that hundreds of rooms would some day be heated by a single fire, yet we see it in every office building or hotel. In equatorial regions there is as much reason for equipping the houses with coolers as there is in temperate regions for equipping them with heaters. In both cases uniformity of temperature is apparently a mistake, for moderate changes from day to day appear to be favorable. Even though a man's work may be out of doors, it seems probable that he would be much stimulated and much better enabled to work hard in the heat if he could sleep in a comparatively cool house. If he lives where the climate is too damp, he would be benefited by having the house relatively dry, just as the northerner in winter apparently ought to have his house more moist than is now his habit. Both need to enjoy the optimum conditions.

A second method of obtaining frequent changes may possibly prove of much importance. Today the seacoast in many regions, for example on the Atlantic shore of America from New York to Boston, is bordered by an almost continuous line of houses. At first people went to these only in the summer. Now many go for week-ends at almost all seasons. Fifty years ago such a thing was almost unknown. Fifty years hence it will probably be many times more prevalent than now. In tropical countries millions of people may not only move to other climates during part of each year, but many may move back and forth from the lowlands to the highlands every few days. Their work may be arranged so that almost every family can spend week-ends in the highlands and the rest of the time in the lowlands. In Sardinia, in order to escape the malaria of summer, many villages are today completely duplicated in the lowlands and highlands, —churches, houses, shops,—everything in duplicate. Moreover, aside from farming, mining, lumbering, and the like, most of the other kinds of work can be located where the climatic conditions are best, a matter which is becoming increasingly easy as our facilities for communication improve. Yet if these are to be within the tropics, the people engaged in them must have an opportunity to obtain the stimulus of changes. Perhaps they will frequently go from their places of work in the highlands to the neighboring lowlands or to the high mountains.

The expense of such a system of having two homes for a large part of the population will doubtless be enormous, but that is relatively unimportant. If the farmers of the tropics were as efficient as those of the temperate zone, one man's labor would produce two or three times as much as in Europe or the United States. If white men can devise a means whereby they can live in the torrid zone and retain approximately the energy which they possess in their own countries, or if they can largely increase the efficiency of the natives, they can afford to spend enormous sums in creating favorable conditions. How we shall go to work in detail cannot yet be determined, but that will easily be discovered. For the present it is enough to see that the hypothesis of climate as a condition of civilization is far from depressing. It holds out hope that the inhabitants of even the most favored parts of the temperate zone may improve their condition. It holds out still more hope that the people of the less favored parts of that zone and of the subtropical zone may be benefited. And it holds out far the greatest hope to those who dwell in the tropical regions which now are the most hopeless.

If our hypothesis is true, man is more closely dependent upon nature than he has realized. A realization of his limitations, however, is the first step toward freedom. In suggesting possible ways of obtaining a new ascendancy over climatic handicaps we have dealt largely with material matters. Bound up with these, and far more important, are great moral issues. We are slowly realizing that character in the broad sense of all that pertains to industry, honesty, purity, intelligence, and strength of will is closely dependent upon the condition of the body. Each influences the other. Neither can be at its best while its companion is dragged down. The climate of many countries seems to be one of the great reasons why idleness, dishonesty, immorality, stupidity, and weakness of will prevail. If we can conquer climate, the whole world will become stronger and nobler.

APPENDIX

APPENDIX

(A) CONTRIBUTIONS TO THE MAP OF CIVILIZATION

1. AMERICANS

J. Barrell, geologist, New Haven, Conn.
P. Bigelow, traveler and author, Malden, N. Y.
I. Bowman, geographer, New York, N. Y.
W. M. Brown, geographer, Providence, R. I.
R. D. Calkins, geographer, Mt. Pleasant, Mich.
J. S. Chandler, missionary, Madura, India.
A. C. Coolidge, historian, Cambridge, Mass.
S. W. Cushing, geographer, Salem, Mass.
L. Farrand, anthropologist, Ithaca, N. Y.
C. W. Furlong, traveler and author, Boston, Mass.
H. Gannett, geographer, Washington, D. C.
E. W. Griffis, traveler and author, Ithaca, N. Y.
A. Hrdlicka, anthropologist, Washington, D. C.
E. H. Hume, physician and missionary, Changsha, China.
E. Huntington, geographer, New Haven, Conn.
M. Jefferson, geographer, Ypsilanti, Mich.
A. G. Keller, anthropologist, New Haven, Conn.
E. F. Merriam, editor, Boston, Mass.
J. H. Potts, missionary, Shanghai, China.
E. Sapir, anthropologist, Ottawa, Canada.
J. R. Smith, economic geographer, New York, N. Y.
E. V. Robinson, economic geographer, Minneapolis, Minn.
W. S. Tower, geographer, New York, N. Y.
R. H. Whitbeck, geographer, Madison, Wisc.
S. W. Zwemer, missionary, Cairo, Egypt.
Anonymous, New York City.

2. British

George Black, Sydney, Australia.
James Bryce, statesman, London.
Leonard Darwin, soldier, London.
T. H. Holdich, soldier, London.
H. H. Johnston, administrator, Arundel, England.
J. S. Keltie, geographer, London.
T. S. Longstaff, geographer, London.
D. Carruthers, explorer, Manningtree, England.

3. Teutons from Continental Europe

S. de Geer, geographer, Stockholm, Sweden.
H. F. Helmholt, historian, München, Germany.
A. Kraemer, ethnographer, Stuttgart, Germany.
Mrs. M. Krug-Genthe, geographer, Chemnitz, Germany.
H. H. Reusch, geologist and geographer, Kristiania, Norway.
H. ten Kate, anthropologist, Holland.
F. von Luschan, anthropologist, Berlin, Germany.
K. F. Sapper, geographer, Strassburg, Germany.

4. Latins (and Other Europeans not Already Classified)

D. n Anoutchine, geographer, Moscow, Russia.
L. Gallois, geographer, Paris, France.
V. Giuffrida-Ruggeri, anthropologist, Naples, Italy.
G. Papillault, anthropologist, Paris, France.
B. y Rospide, geographer, Madrid, Spain.
G. Sergi, anthropologist, Rome, Italy.
S. Telles, geographer, Lisbon, Portugal.

5. Asiatics

Katsuro Haro, Imperial University, Kyoto, Japan.
Inazo Nitobe, Imperial University, Tokyo, Japan.
Naomasa Yamasaki, Imperial University, Tokyo, Japan.
Jeme Tien-yow, Shanghai, China.
Wang Ching-chun, Tientsin, China.
Wu Ting-fang, Shanghai, China.

SUMMARY OF CONTRIBUTORS

	Letters sent	Replies	Contributors
Americans	64	57	25
British	43	38	8
Teutonic Europeans	42	23	8
Latin Europeans	27	8	6
Asiatics	21	8	6
Non-Teutonic and non-Latin Europeans .	10	2	1
Latin Americans	6	2	0
Total	213	138	54

(B) THE RELATIVE CIVILIZATION OF THE COUNTRIES OF THE WORLD

The following list contains the final results of the classification made by the preceding contributors. In examining this the reader should remember that the division into regions and the grouping of the regions according to race make no claim to perfection. Convenience in obtaining units small enough to give a detailed map and yet to be known to people of many races has been the primary object in dividing the different countries of the world into smaller sections. The sections are small in Europe and the United States because these regions are well known, and large in Siberia because very few people can distinguish sharply between different parts of that country. In a few cases I have added minor divisions such as southern Alaska, with a special object not connected with the present book. The grouping by races in Europe, Asia, and North America has been guided also by motives of convenience. Such places as Ireland, Asia Minor, Bulgaria, Baluchistan, and others might properly be placed in other groups as well as in those where I have put them. The reader can easily rearrange for himself. My purpose has been merely to make a convenient classification for our immediate purpose without respect to its applicability elsewhere.

A CLASSIFICATION OF THE REGIONS OF THE WORLD ACCORDING TO THEIR CIVILIZATION.

BY FIFTY CONTRIBUTORS

NAME OF REGION	RANK ACCORDING TO VARIOUS GROUPS OF CONTRIBUTORS					
	25 AMERICANS	7 BRITISH	6 TEUTONS	7 LATINS	5 ASIATICS	All CONTRIBUTORS
EUROPE						
Teutonic Regions						
1. England and Wales	100	100	100	100	100	100
2. Northwestern Germany (from Berlin north and west)	98	99	100	99	100	99
3. Central Germany (Saxony and westward to Luxemburg on the south and northwestward to Hanover)	98	99	100	100	98	99
4. Scotland	98	99	98	96	100	98
5. Denmark	97	97	100	100	94	98
6. Holland	99	95	98	100	96	98
7. Switzerland	96	97	98	100	96	97
8. Southern Germany (Bavaria and Rhine Basin to Bingen)	94	92	98	100	98	97
9. Sweden	94	96	98	100	92	96
10. Belgium	94	90	94	100	96	95
11. Austria Proper	90	93	92	96	98	94
12. Norway	92	92	93	100	92	94
13. Southeastern Germany (south of Berlin and east of longitude 14° E.)	91	96	95	90	92	93
14. Northeastern Germany (north and east of Berlin, but not including that city)	90	92	96	86	94	92
15. Ireland	81	90	70	80	90	82
16. Austrian Alps	74	82	90	77	74	79

NAME OF REGION	25 AMERICANS	7 BRITISH	6 TEUTONS	7 LATINS	5 ASIATICS	All CONTRIBUTORS
Latin Regions						
17. Northern France (Paris Basin, etc.)	100	99	96	100	100	99
18. Northern Italy (as far south as Florence)	92	92	90	100	94	94
19. Southern France and Rhone Valley	91	85	87	97	98	91
20. Western France (Brittany and southward)	87	92	90	90	90	90
21. Central Italy (not including Florence, but including Naples)	81	83	85	90	84	85
22. Central France (mountainous portions)	82	86	86	87	84	85
23. Central Spain (south of latitude 42° and north of the Guadalquiver River)	68	76	70	79	82	75
24. Northern Spain (approximately north of latitude 42°)	71	75	72	77	72	73
25. Roumania	68	75	75	76	68	72
26. Greece	71	72	70	76	72	72
27. Southern Italy and Sicily (not including Naples)	66	72	68	70	82	72
28. Southeastern Spain (Andalusia, Granada, etc.)	66	75	64	71	82	72
29. Portugal	68	66	67	73	76	70
30. Corsica and Sardinia	61	70	66	65	56	64
31. Albania and Montenegro	56	63	50	57	52	56
Slavic Regions						
32. Baltic Provinces	78	88	82	83	78	82
33. Bohemia and Moravia	77	80	84	86	76	81
34. Central Russia (south of St. Petersburg and Perm, west of the Volga, and north of latitude 50°)	62	75	74	73	82	73
35. Poland	72	76	66	74	68	71
36. Transcarpathian Austria (Galicia, etc.)	66	72	64	70	70	68

RANK ACCORDING TO VARIOUS GROUPS OF CONTRIBUTORS

NAME OF REGION	25 AMERICANS	7 BRITISH	6 TEUTONS	7 LATINS	5 ASIATICS	All CONTRIBUTORS
Slavic Regions (Continued)						
37. Servia	64	70	62	70	66	66
38. Adriatic Provinces (including Bosnia)	61	70	67	69	60	65
39. Southern Russia (west of longitude 40° and south of latitude 50°)	64	70	64	59	60	63
40. Northeastern Russia (north of latitude 60°)	46	64	44	53	46	51
41. Southeastern Russia (Caspian Steppes)	47	54	42	48	40	46
Uralo-Altaic Regions						
42. Hungary	77	80	76	88	90	82
43. Southern Finland	76	83	94	83	66	80
44. Bulgaria	63	72	57	67	70	66
45. Turkey in Europe (including portions ceded to Bulgaria in 1913)	57	60	59	61	66	61
46. Lapland	34	37	37	20	18	29
NORTH AMERICA						
Teutonic Regions						
47. North Atlantic States (New York, Pennsylvania, and New Jersey)	99	100	100	100	100	100
48. New England	100	100	100	97	100	99
49. Lake States (Wisconsin, Michigan, Illinois, Indiana, and Ohio)	98	93	90	91	100	95
50. California	95	87	90	89	90	90
51. Quebec	85	89	88	90	86	88
52. Washington and Oregon	94	89	92	79	84	88

NAME OF REGION	25 AMERICANS	7 BRITISH	6 TEUTONS	7 LATINS	5 ASIATICS	All CONTRIBUTORS
Teutonic Regions (Continued)						
53. Ontario	94	90	94	76	82	87
54. Central Prairie States (Kansas, Missouri, Oklahoma, and Arkansas)	88	84	84	83	88	86
55. Northern Prairie States (North Dakota, South Dakota, Minnesota, Nebraska, and Iowa)	93	81	80	77	84	83
56. Northern Rocky Mountain States (Montana, Idaho, and Wyoming)	86	81	90	76	76	82
57. Maritime Provinces of Canada	88	77	82	84	74	81
58. British Columbia	84	89	76	63	74	77
59. Manitoba	88	76	75	77	58	75
60. Southern Rocky Mountain States (Utah, Colorado, Nevada, Arizona, and New Mexico)	82	64	75	70	74	73
61. Newfoundland	76	80	84	52	68	72
62. Alberta and Saskatchewan	86	80	70	44	40	64
63. Southern Greenland and Iceland	47	54	83	34	32	50
64. Southern Alaska	56	47	40	37	38	44
Teutonic and Negro Regions						
65. Central Atlantic States (North Carolina, Virginia, Maryland, and Delaware)	92	87	83	94	96	90
66. Central Appalachian States (West Virginia, Kentucky, and Tennessee)	84	81	90	89	94	88
67. South Atlantic States (Florida, Georgia, and South Carolina)	77	78	72	80	86	79

NAME OF REGION	RANK ACCORDING TO VARIOUS GROUPS OF CONTRIBUTORS					
	25 AMERICANS	7 BRITISH	6 TEUTONS	7 LATINS	5 ASIATICS	All CONTRIBUTORS
Teutonic and Negro Regions (Continued)						
68. Central Gulf States (Louisiana, Mississippi, and Alabama)	76	71	72	87	82	78
69. Texas	83	73	73	69	82	76
Latin American Regions						
70. Central Mexico (including Mexico City and the main highlands)	53	64	68	70	64	64
71. Northern Mexico to latitude 23°	50	59	48	46	54	51
72. West Indies	46	57	50	60	42	51
73. Central America	48	49	44	50	52	48
74. Southern Mexico	40	43	48	44	52	45
Indian Regions						
75. Sparsely populated forest regions of Alaska, northern Canada, and Labrador	27	34	15	13	18	21
76. Northern coasts of North America (Eskimo regions including most of Greenland)	16	23	18	11	18	17
ASIA						
Sino-Japanese and Mongoloid Regions						
77. Southern Japan (south of latitude 38°)	85	80	75	77	96	83
78. Central China (Yangtse Basin)	63	53	62	60	90	66
79. Northern China (Hoang-Ho Basin)	65	53	62	60	88	65
80. Northern Japan (Yezo and northern quarter of Hondo)	60	61	67	41	78	62
81. Southern China (Si-kiang Basin)	58	44	60	55	84	60

NAME OF REGION	25 AMERICANS	7 BRITISH	6 TEUTONS	7 LATINS	5 ASIATICS	All CONTRIBUTORS
Sino-Japanese and Mongoloid Regions (Continued)						
82. Southeastern China (coast provinces of Chekiang and Fukien)	59	49	56	51	86	60
83. Southwestern China (Yun-nan and Kweichou)	48	41	56	40	60	49
84. Southern Manchuria	51	37	52	42	58	48
85. Siam	44	41	48	46	58	47
86. Korea	46	39	50	43	46	45
87. Burma (south of latitude 20°)	42	41	42	42	46	43
88. Indo-China (south of latitude 15°)	36	29	42	46	44	39
89. Northern Manchuria	40	37	38	29	42	37
90. Indo-China (north of latitude 15°)	34	30	38	38	40	36
91. Burma (north of latitude 20°)	38	40	40	23	36	35
92. Tibet	30	39	32	30	32	33
93. Mongolia	32	36	28	29	32	31
Indo-European Regions						
94. United Provinces of India, Punjab, and northwest frontier	52	59	50	46	60	53
95. Central Siberia (agricultural region along the Siberian railroad to Lake Baikal)	55	54	57	54	44	53
96. Bengal and Assam	51	59	48	43	58	52
97. Caucasus	51	56	52	47	48	51
98. Central India (the peninsula from about latitude 23° to 15°)	49	54	48	41	60	51
99. Rajputana and Sind	46	57	45	49	30	45
100. Armenian Plateau	50	49	47	39	42	45

NAME OF REGION	25 AMERICANS	7 BRITISH	6 TEUTONS	7 LATINS	5 ASIATICS	All CONTRIBUTORS
Indo-European Regions (Continued)						
101. Northwestern Persia	43	50	44	42	48	45
102. Southern India (south of latitude 15°)	43	46	48	36	50	44
103. Northeastern Persia	40	41	40	36	40	40
104. Southern and eastern Persia	39	44	36	40	36	39
105. Afghanistan	34	43	36	37	40	38
106. Baluchistan	34	31	32	34	32	33
Turanian or Uralo-Altaic Regions						
107. Asia Minor	50	53	42	49	42	47
108. Russian Turkestan	40	41	39	43	34	39
109. Amur region of Siberia	41	39	35	39	26	36
110. Chinese Turkestan	34	36	32	26	30	31
111. Steppes of southern Siberia	38	37	28	29	22	31
112. Mountains of eastern Siberia (northeast of Lake Baikal to Kamchatka)	24	24	17	22	22	22
113. Transcaspian Desert	29	31	15	21	12	22
114. Forests of northern Siberia	22	23	10	14	12	16
115. Tundras of Arctic coast of Asia	17	16	10	11	10	13
Semitic Regions						
116. Syria	48	47	47	51	48	48
117. Mesopotamia	38	40	34	37	36	37
118. Yemen	32	39	32	30	26	32
119. Oman	31	27	28	29	20	29
120. Syrian Desert	28	30	30	21	10	24
121. Arabian Desert	25	37	18	17	14	22

NAME OF REGION	25 AMERICANS	7 BRITISH	6 TEUTONS	7 LATINS	5 ASIATICS	All CONTRIBUTORS
East India and Malaysian Regions						
122. Java	36	43	52	53	38	44
123. Philippines	40	33	37	39	44	39
124. Malay Peninsula	35	49	38	26	28	35
125. Formosa	30	27	30	33	38	32
126. Sumatra	29	31	32	23	24	28
127. Celebes	23	26	30	21	24	25
128. Borneo	21	27	22	16	18	21
129. New Guinea	14	16	10	12	26	15
AUSTRALIA						
130. New South Wales	84	93	88	83	82	86
131. Victoria and Tasmania	83	87	94	80	78	84
132. New Zealand	88	94	93	82	80	87
133. Southern Queensland	72	76	84	79	78	78
134. South Australia	74	87	70	74	64	74
135. West Australia (south of latitude 30°)	66	74	60	51	68	62
136. West Australia (from latitude 23 1-2° to 30° S.)	55	54	68	50	58	59
137. Northern Queensland	46	51	58	40	62	52
138. North Territory of South Australia	44	46	30	30	54	41
139. West Australia (north of latitude 20° S.)	49	41	28	26	50	39
AFRICA						
140. Cape Colony	73	86	67	70	64	72
141. Transvaal and Orange River Colony	65	67	57	64	54	62

NAME OF REGION

RANK ACCORDING TO VARIOUS GROUPS OF CONTRIBUTORS

NAME OF REGION	25 AMERICAN	7 BRITISH	6 TEUTONS	7 LATINS	5 ASIATICS	ALL CONTRIBUTORS
AFRICA (*Continued*)						
142. Natal	52	82	65	59	60	62
143. Egypt	44	54	60	62	64	57
144. Algeria and Tunisia	48	54	53	63	50	54
145. South Rhodesia	47	59	37	40	44	45
146. Morocco	37	40	33	39	38	38
147. Tripoli	36	41	36	31	32	35
148. Abyssinia	34	43	46	27	32	35
149. Madagascar	30	29	33	38	36	33
150. North Rhodesia	39	48	24	24	18	31
151. British East Africa (except Uganda)	26	34	30	26	20	27
152. Uganda	27	36	28	24	20	26
153. Egyptian Sudan	26	37	22	23	22	26
154. German East Africa	20	27	35	26	20	26
155. German Southwest Africa	28	27	24	27	20	25
156. Sudan	23	29	27	23	22	25
157. North coast of the Gulf of Guinea (south of latitude 10°)	15	20	14	25	34	22
158. Angola	26	23	17	24	18	22
159. Somaliland	19	26	18	21	18	20
160. Portuguese East Africa	20	21	15	23	20	20
161. Kamerun, French Congo, and Belgian Congo	19	20	16	24	20	20
162. Sahara	23	21	18	18	14	19
163. Kalahari Desert	16	14	10	10	10	12

NAME OF REGION	25 Americans	7 British	6 Teutons	7 Latins	5 Asiatics	All Contributors
SOUTH AMERICA						
164. Argentina (between 30° and 40° S.)	68	67	65	74	80	71
165. Central Chili	62	63	72	63	50	62
166. Brazil (south of latitude 20°)	64	48	58	63	66	60
167. Argentina (north of latitude 30°)	57	53	53	52	66	56
168. Uruguay	61	43	48	64	48	53
169. Paraguay	46	40	50	51	50	48
170. Mountainous part of Peru	41 (45)	41 (49)	40 (45)	43 (46)	34 (46)	40 (46)
171. Mountainous part of Colombia	46 (49)	40 (44)	34 (40)	47 (48)	34 (46)	40 (45)
172. Ecuador	44	37	38	53	50	44
173. Mountainous part of Bolivia	40 (43)	40 (47)	38 (38)	40 (40)	36 (46)	30 (43)
174. Mountainous eastern region of Brazil (north of latitude 20°)	41	33	40	41	48	41
175. Mountainous part of Venezuela	41 (46)	27 (33)	33 (35)	41 (43)	34 (46)	35 (41)
176. Guiana (French, British, and Dutch)	30	34	33	34	38	34
177. Northern desert portion of Chili	31	37	32	19	34	31
178. Southern Chili (including Tierra del Fuego)	29	30	28	29	36	30
179. Eastern lowlands of Colombia	26 (24)	39 (34)	26 (20)	23 (22)	46 (34)	32 (28)
180. Eastern lowlands of Bolivia	28 (24)	34 (27)	18 (18)	26 (26)	44 (34)	30 (26)
181. Matto Grosso (central region of Brazil south of latitude 10°)	30	24	22	19	38	25
182. Patagonia	24	24	34	20	22	25
183. Eastern lowlands of Peru	26 (22)	37 (30)	23 (18)	24 (21)	44 (32)	31 (25)
184. Orinoco lowlands of Venezuela	29 (25)	26 (20)	26 (20)	24 (23)	42 (30)	28 (24)
185. Amazon Basin	20	20	13	20	40	23

NOTE.—In the Andean countries of South America, except Chili, the most advanced districts are found in the highlands. The eastern lowland, which is heavily forested or else covered with savannah, contains only a scanty population composed almost wholly of Indians. The way in which each of these countries is divided into a "mountainous part," and an "eastern lowland" seems to have created confusion in the minds of many contributors. They apparently thought of the lofty mountains, and not of the plateau as was intended. Hence they rank the lowlands higher than the mountains. As a matter of fact, the lowlands are in almost the same condition as the Amazon Basin (No. 185). Therefore, after the figures for Venezuela, Colombia, Peru, and Bolivia, I have added in parentheses a series of numbers indicating the rank of these countries if in each case the higher value in any given classification is taken as intended for the more advanced portion, that is, the highlands. The numbers in parentheses are probably nearer right than the others, and have been used in preparing the map of civilization, Figure 45.

CLASSIFICATION OF THE STATES AND PROVINCES OF THE UNITED STATES AND CANADA

UNITED STATES	Rank per first 10 contributors A.	Rank per second 10 contributors B.	Difference between A and B C.	Final rank per 23 contributors D.
1. Alabama	2.1	1.9	0.2	2.0
2. Arizona	1.2	1.7	0.5	1.6
3. Arkansas	2.3	1.9	0.4	2.0
4. California	5.3	4.8	0.5	5.1
5. Colorado	4.1	4.1	0.0	4.2
6. Connecticut and Rhode Island	5.9	5.8	0.1	5.8
7. Delaware	4.0	3.9	0.1	4.0
8. Florida	2.0	2.0	0.0	2.0
9. Georgia	2.1	2.5	0.4	2.5
10. Idaho	2.5	2.5	0.0	2.6
11. Illinois	5.7	5.7	0.0	5.7
12. Indiana	5.3	5.6	0.3	5.4
13. Iowa	5.2	5.1	0.1	5.0
14. Kansas	4.8	5.0	0.2	4.9
15. Kentucky	3.2	3.3	0.1	3.3
16. Louisiana	2.0	1.7	0.3	2.0
17. Maine	4.9	5.0	0.1	4.8
18. Maryland	4.1	4.0	0.1	4.1
19. Massachusetts	6.0	6.0	0.0	6.0
20. Michigan	5.4	5.6	0.2	5.4
21. Minnesota	5.1	5.4	0.3	5.2
22. Mississippi	1.7	1.4	0.3	1.7
23. Missouri	3.7	3.5	0.2	3.7
24. Montana	3.0	3.2	0.2	3.1
25. Nebraska	4.4	4.3	0.1	4.4
26. Nevada	1.3	1.9	0.6	1.7
27. New Hampshire	4.8	4.8	0.0	4.6
28. New Jersey	5.7	5.4	0.3	5.6
29. New Mexico	1.6	1.5	0.1	1.6
30. New York	5.9	5.7	0.2	5.8
31. North Carolina	3.1	2.8	0.3	2.9
32. North Dakota	3.8	3.8	0.0	3.8
33. Ohio	5.8	5.9	0.1	5.8

		Rank per first 10 contributors	Rank per second 10 contributors	Difference between A and B	Final rank per 23 contributors
	UNITED STATES (*Continued*)	A.	B.	C.	D.
34.	Oklahoma	3.7	3.1	0.6	3.4
35.	Oregon	4.6	5.0	0.4	4.9
36.	Pennsylvania	5.6	5.5	0.1	5.6
37.	South Carolina	1.6	2.2	0.6	1.9
38.	South Dakota	3.8	3.7	0.1	3.7
39.	Tennessee	3.0	2.6	0.4	2.8
40.	Eastern Texas	3.1	2.8	0.3	3.1
41.	Western Texas	1.8	1.6	0.2	1.9
42.	Utah	2.4	2.3	0.1	2.5
43.	Vermont	4.4	4.7	0.3	4.4
44.	Virginia	3.8	3.8	0.0	3.9
45.	Washington	4.8	5.0	0.2	5.0
46.	West Virginia	2.5	2.9	0.4	2.8
47.	Wisconsin	5.4	5.8	0.4	5.6
48.	Wyoming	2.9	2.6	0.3	2.8
49.	Southern Alaska	1.5	1.2	0.3	1.5
	CANADA	A.	B.	C.	D.
50.	Newfoundland	1.9	2.0	0.1	2.0
51.	Prince Edward's Island . .	2.6	2.1	0.5	2.5
52.	Nova Scotia	3.8	3.4	0.4	3.6
53.	New Brunswick	3.3	2.9	0.4	3.2
*54.	Quebec, east of longitude 72° 30′	3.1	2.6	0.5	3.0
*55.	Quebec, west of longitude 72° 30′	4.1	2.6	1.5	3.4
56.	Ontario, east of Lake Huron .	5.4	4.7	0.7	5.1
57.	Ontario, north of Lakes Huron and Superior	3.0	2.0	1.0	2.6

* In this case some contributors may have misunderstood what was intended. At least, several of them placed the part of Quebec containing Montreal lower than the sparsely populated portion from Quebec eastward. If each contributor's higher figure be taken as meant for the part of the province containing Montreal, the final numbers in column D become East Quebec 2.5 and West Quebec 3.9. I am strongly inclined to think that many of the Canadian figures are too low, because all but one of the contributors were from the United States. For this reason Canada is not included in Figure 43.

	Rank per first 10 contributors	Rank per second 10 contributors	Difference between A and B	Final rank per 23 contributors
CANADA (*Continued*)	A.	B.	C.	D.
58. Manitoba	3.5	4.4	0.9	4.0
59. Saskatchewan, southern half, i.e., south of latitude 55° . . .	3.0	3.0	0.0	3.1
60. Saskatchewan, northern half, i.e., north of latitude 55° .	1.3	2.0	0.7	1.7
61. Alberta, southern half . . .	2.9	2.9	0.0	3.1
62. Alberta, northern half . . .	1.7	1.7	0.0	1.8
63. British Columbia, southern half	3.8	3.4	0.4	3.8
64. British Columbia, northern half	1.3	1.9	0.6	1.7

(C) EFFICIENCY AND MEAN TEMPERATURE

A TABLE BASED ON THE WORK OF 310 MEN AND 196 WOMEN
AT NEW HAVEN, NEW BRITAIN, AND BRIDGE-
PORT IN 1910-1913

TEMP. C.	EFFICIENCY	TEMP. C.	EFFICIENCY
—35°	92.8%	—21°	93.5%
—34°	92.8%	—20°	93.6%
—33°	92.9%	—19°	93.6%
—32°	92.9%	—18°	93.7%
—31°	92.9%	—17°	93.8%
—30°	93.0%	—16°	93.9%
—29°	93.0%	—15°	94.0%
—28°	93.1%	—14°	94.1%
—27°	93.1%	—13°	94.2%
—26°	93.2%	—12°	94.3%
—25°	93.3%	—11°	94.4%
—24°	93.3%	—10°	94.5%
—23°	93.4%	— 9°	94.6%
—22°	93.4%	— 8°	94.7%

TEMP. C.	EFFICIENCY	TEMP. C.	EFFICIENCY
— 7°	94.9%	15°	100.0%
— 6°	95.1%	16°	99.9%
— 5°	95.3%	17°	99.7%
— 4°	95.5%	18°	99.3%
— 3°	95.7%	19°	98.9%
— 2°	95.9%	20°	98.4%
— 1°	96.2%	21°	97.9%
0°	96.5%	22°	97.4%
1°	96.8%	23°	96.9%
2°	97.1%	24°	96.4%
3°	97.4%	25°	95.9%
4°	97.7%	26°	95.3%
5°	98.0%	27°	94.6%
6°	98.3%	28°	93.9%
7°	98.6%	29°	93.2%
8°	98.8%	30°	92.5%
9°	99.0%	31°	91.7%
10°	99.2%	32°	90.5%
11°	99.4%	33°	89.0%
12°	99.6%	34°	86.8%
13°	99.8%	35°	84.3%
14°	99.9%	36°	81.0%

INDEX

INDEX